THE MINI-ATLAS OF SNAKES OF THE WORLD

by John Coborn

t.f.h.

© 1994 by T.F.H. Publications, Inc. USA

Distributed in the UNITED STATES to the Pet Trade by T.F.H. Publications, Inc., One T.F.H. Plaza, Neptune City, NJ 07753; distributed in the UNITED STATES to the Bookstore and Library Trade by National Book Network, Inc. 4720 Boston Way, Lanham MD 20706; in CANADA to the Pet Trade by H & L Pet Supplies Inc., 27 Kingston Crescent, Kitchener, Ontario N2B 2T6; Rolf C. Hagen Ltd., 3225 Sartelon Street, Montreal 382 Quebec; in CANADA to the Book Trade by Macmillan of Canada (A Division of Canada Publishing Corporation), 164 Commander Boulevard, Agincourt, Ontario M1S 3C7; in ENGLAND by T.F.H. Publications, PO Box 15, Waterlooville PO7 6BQ; in AUSTRALIA AND THE SOUTH PACIFIC by T.F.H. (Australia), Pty. Ltd., Box 149, Brookvale 2100 N.S.W., Australia; in NEW ZEALAND by Brooklands Aquarium Ltd. 5 McGiven Drive, New Plymouth, RD1 New Zealand; in Japan by T.F.H. Publications, Japan—Jiro Tsuda, 10-12-3 Ohjidai, Sakura, Chiba 285, Japan; in SOUTH AFRICA by Multipet Pty. Ltd., P.O. Box 35347, Northway, 4065, South Africa. Published by T.F.H. Publications, Inc.

MANUFACTURED IN THE UNITED STATES OF AMERICA
BY T.F.H. PUBLICATIONS, INC.

TABLE OF CONTENTS

Introduction 4
Snakes of the World 6
Listing of Snake Genera 26
Photographers and Illustrators 29
Key to the Symbols 30
Pictorial Section 32

Family Typhlopidae 33	Subfamily Natricinae 253
Family Anomalepidae 34	Subfamily Homalopsinae 309
Family Leptotyphlopidae 35	Subfamily Boiginae 312
Family Aniliidae 36	Subfamily Dipsadinae 392
Family Boidae 38	Subfamily Pareinae 394
Subfamily Loxoceminae 38	Subfamily Aparallactinae 396
Subfamily Pythoninae 39	Family Elapidae 398
Subfamily Boinae 50	Family Hydrophiidae 451
Subfamily Erycinae 66	Subfamily Laticaudinae 451
Family Xenopeltidae 73	Subfamily Hydrophiinae 453
Family Colubridae 74	Family Viperidae 462
Subfamily Xenodontinae 74	Subfamily Azemiophinae 462
Subfamily Colubrinae 89	Subfamily Viperinae 463
Subfamily Dasypeltinae 235	Subfamily Causinae 510
Subfamily Lycodontinae 236	Family Crotalidae 513

Snake Natural History 577
Classification 584
Biology of Reptiles 588
The Terrarium 615
Nutrition 634
General Care 647
Hygiene and Disease 659
Reproduction and Captive Breeding 670
The Human Connection 688
Snakes in Medicine 698
Suggested Reading 709
Index 711

Introduction

Though mini by name, the *Mini-Atlas of Snakes of the World* is certainly not mini by nature. It contains a lot of the information included in my original *Atlas of Snakes of the World* (also published by TFH) plus many new photographs never before published. The illustrations are indeed lavish, but this book has smaller dimensions than its larger brother, weighs less, and is thus more convenient and economical to the majority of snake enthusiasts. The original idea of the *Atlas* was to bring together a vast amount of material about snake habits and habitats, to apply this to their requirements in captivity, and to illustrate as many species as possible. In this smaller guide, of course, much of this has been condensed, but the original intentions are still intact. Neither *Atlas* is meant to be a formal identification guide, for example, but they nevertheless depict many animals that have not been seen before by the average reader.

Due to the plethora of conflicting literature with regard to snake taxonomy, it is very difficult, even impossible, to come up with a system which will be acceptable to all workers in this discipline. However, I have endeavored to ascertain a sensible median in this respect through the information available to me, and thus any obvious errors will be mine and mine alone. In a corner of science such as systematics and taxonomy, where there are dozens of fiercely opposing opinions, it is to be expected that some people will oppose my viewpoints. I therefore offer my apologies in advance to anybody who falls into this category.

Our techniques in modern reptile husbandry are necessarily based on our knowledge of wild habitats and behavior of individual species, coupled with the collective published experiences of generations of herpetologists, both professional and amateur. Indeed, contributions by amateurs should not be underestimated, and the keepers of even the simplest of terraria have often been able to contribute new and exciting snippets of information to our list of herpetological facts and figures.

Like the original *Atlas*, I have also prepared this book in the hope that it will alleviate some of the difficulties experienced by professional and amateur herpetologists who seek any and all information on snakes and their care. The book is also intended to raise the enthusiasm of aspiring snake keepers and hopefully to lead them into the many years of interest and pleasure which the serious serpentophile will be bound to experience.

John Coborn
Nanango, Queensland
Australia

The snakes of the world have fascinated man for centuries. Shown here are two illustrations done in the mid-1800's. **Above:** the Timber Rattlesnake, *Crotalus horridus*. **Below:** the Black Rat Snake, known at that time as the Pilot Black Snake, *Elaphe obsoleta obsoleta*.

Snakes of the World

HOW TO USE THIS SECTION

Snakes of the world are arranged systematically in family and subfamily rank. Unlike the original *Atlas*, specific accounts of the genera are not included although there are brief descriptions concerning the genera of each family and/or subfamily. The arrangement follows the same order of the original *Atlas*. The photos in the pictorial section are naturally arranged in the same systematic order as the accounts in the text, but it should be noted that not every family or subfamily is represented by a photo or series of photos. An attempt has been made to be as complete with this section as possible, but obviously such a task in too monumental for a book of this nature.

As in the *Atlas*, as attempt has been made to follow spellings and dates in Williams and Wallach, 1989, when possible; this has resulted in several changes from the information generally accepted in earlier terrarium works. Common English names have been used only in the index, and will probably be of great relief to the scores of hobbyists who have long since avoided the sometimes tedious process of trying to learn the scientific monikers.

Also in the original *Atlas* were general data on geographical distribution, habitats, habits, general appearance, and care. In this smaller book, much of this information has been greatly condensed, but much of it still exists nevertheless.

FAMILY TYPHLOPIDAE—BLIND SNAKES

The Typhlopidae and the following two families of burrowing snakes are placed together in the infraorder Scolecophidia. The Typhlopidae contains some 240 species in at least 3 genera. They are most abundant in Africa and Asia. The rostral scale overhangs the mouth to form a shovel-like burrowing structure. The eyes are covered by enlarged scales and the teeth occur only in the upper jaw. The short tail ends with a horn-like scale. Most species are oviparous.

FAMILY ANOMALEPIDAE—AMERICAN BLIND SNAKES

Generally similar to Typhlopidae but some species possess a single tooth in the lower jaw. Family contains four genera and some 20 species native to Central and South America

FAMILY LEPTOTYPHLOPIDAE—WORM SNAKES

Though under revision, the family is generally accepted to contain two genera and about

41 species occurring in the Americas, Africa, and Asia, as far as India. May be found in arid areas to rainforest. Occur in the nests of ants or termites, or in the root balls of epiphytes. Contains the world's smallest snakes, for example, *Leptotyphlops tesselatus*, at 8 cm (3.25 in). Very slender and wormlike; round in section. No teeth in upper jaw; feed largely on termites or ants, their larvae and pupae. Most species suck out the contents of insect bodies and discard skin.

FAMILY ANILIIDAE—CYLINDER SNAKES

This and the following four families of snakes are usually grouped together in the infraorder Henophidia. Almost all of the species possess a vestigial pelvic girdle that, in many of them, visibly manifests itself as cloacal spurs. Aniliidae contains 3 genera and about 10 species though *Anomochilus* and *Cylindrophis* are assigned to the family Uropeltidae by some specialists. All are fossorial to a greater or lesser degree. All are ovoviviparous and possess cloacal spurs. Food consists mainly of small amphibians and reptiles. In captivity, cylinder snakes require a heated tropical terrarium with a deep, well-drained substrate for burrowing. Narrow glass

Leptotyphlops dulcis: Leptotyphlopidae

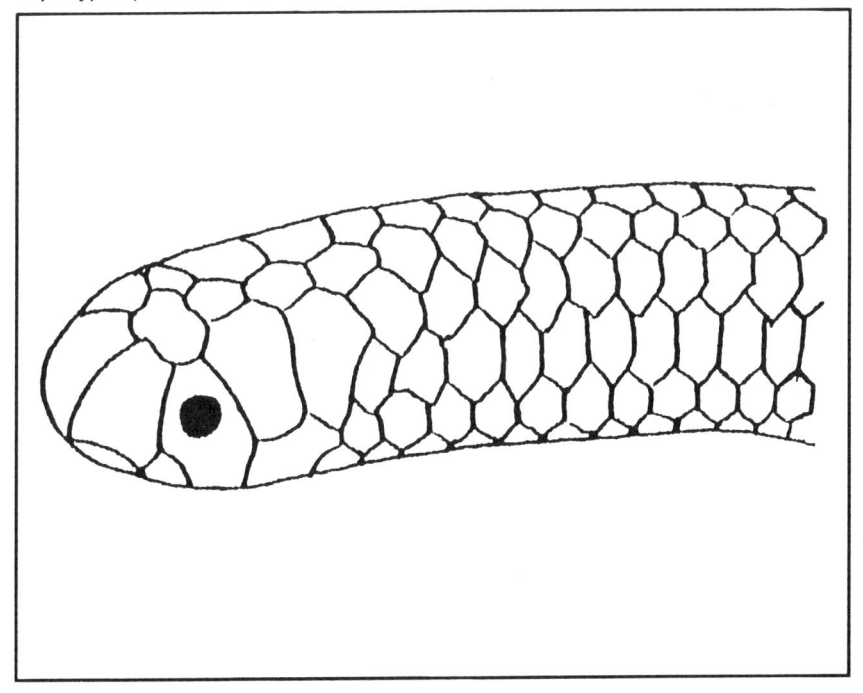

containers provide better possibilities for observation.

FAMILY ACROCHORDIDAE— WART, FILE OR ELEPHANT'S TRUNK SNAKES

A family containing a single genus with three species found from India through S.E. Asia, to New Guinea, the Solomon Islands, and northern Australia. Almost totally aquatic and may occur in fresh or brackish water, sometimes in sea water. Length to 2.5 m (8 ft). The head and body are covered with small, finely-keeled and pointed scales with a rasp-like texture. There are no enlarged belly scales. The whole skin is somewhat flabby, hanging loosely on the body. The small eyes and valved nostrils are arranged on the top of the broad head. The tail is prehensile. All species are ovoviviparous.

File snakes should be kept in large, heated aquaria with underwater landscaping (rocks and branches) so that the reptiles have something to hang onto with their tails. A little salt added to the water will be beneficial. Feed on live fish. Newly captured specimens are often reluctant to feed so force-feeding may be necessary.

FAMILY BOIDAE—PYTHONS AND BOAS

With 5 subfamilies, 23 genera, and about 80 species, distributed throughout the tropics. The size range is enormous, from 45 cm (18 in) to 10 m (33 ft) and the group includes the world's largest snake. All possess a vestigial pelvic girdle usually manifesting itself as a pair of short, claw-like spurs situated on either side of the vent. The jaws are furnished with many strong, recurved teeth. Prey is overcome by constriction. They may be oviparous or ovoviviparous and some of the former exhibit maternal care (brooding) to a greater or lesser extent.

Some authorities would prefer to divide the Boidae into smaller families rather than use subfamilial status.

In view of their size, beauty, and their readiness to become tame, many species in this family are extremely popular terrarium subjects. However, some of the larger species will soon outgrow their accommodations, so unless one has the space to provide a relatively large terrarium, one should perhaps elect to keep smaller species. Additionally, local laws in some areas restrict the keeping of boids by private individuals.

Subfamily Loxoceminae—Mexican Dwarf Pythons

Contains only a single genus.

Subfamily Pythoninae—Pythons

This subfamily contains 6 genera and 22 species found in tropical Africa, Asia, and Australasia. Many are the giants of the snake world and require voluminous accommodations, including a large, heated water container for bathing.

8

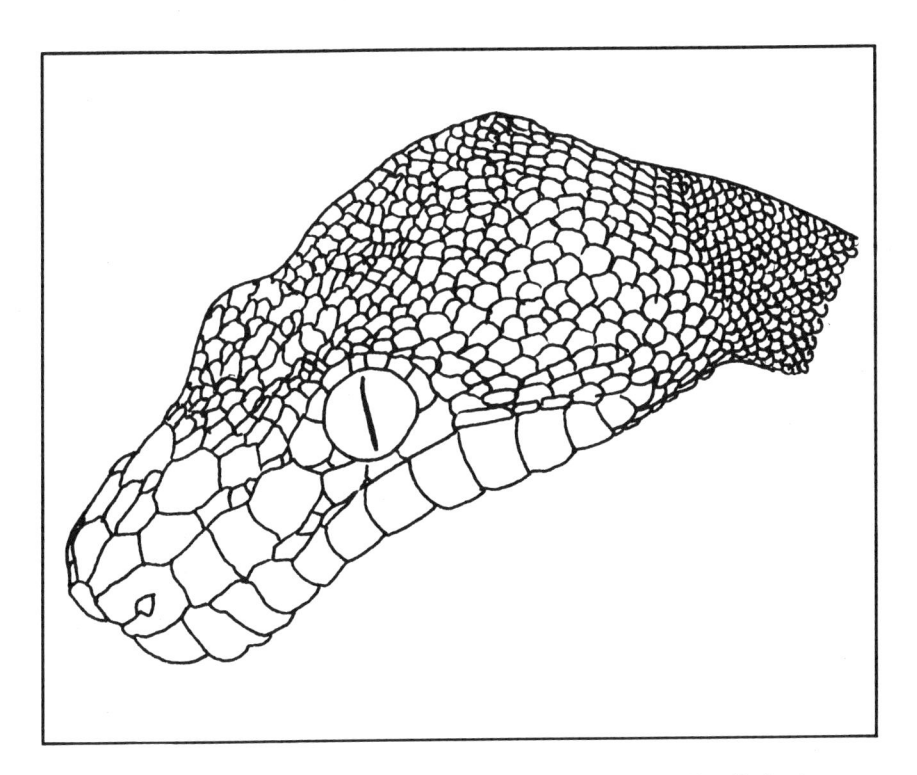

Above: *Corallus caninus:* Boidae: Boinae. **Below:** *Python boeleni:* Boidae: Pythoninae.

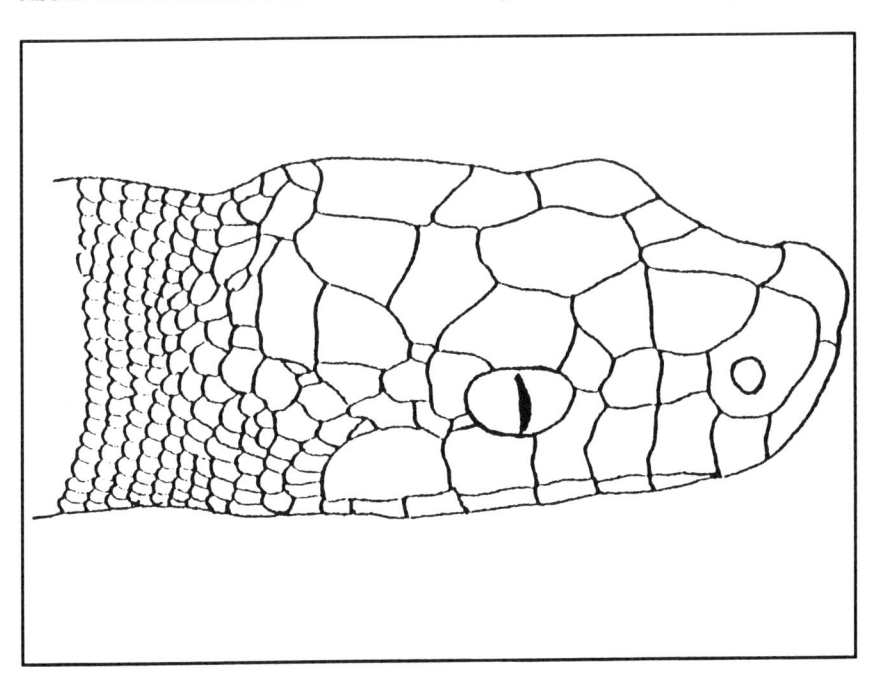

Subfamily Boinae—Typical Boas

This subfamily contains 11 genera and some 41 species mainly found in the New World but also in Madagascar, New Guinea and some Pacific Islands. All are ovoviviparous.

Subfamily Erycinae—Sand Boas

Contains three genera and about 14 species of sand, rubber and Pacific boas. Taxonomy under revision.

Subfamily Bolyeriinae—Round Island Boas

Currently contains two monotypic genera. One, *Bolyeria*, was described by Gray in 1842. Specimens were only found on Round Island, off Mauritius in the Indian Ocean, but none have been seen since 1975 and thus the genus is believed to be extinct. The second genus, *Casarea* (also described by Gray, 1842), is also endemic to this tiny tropical island.

FAMILY UROPELTIDAE—SHIELDTAIL SNAKES

A family of mainly small (20-50 cm—8-20 in) burrowing snakes containing 8 genera and some 44 species found only in the southern part of India and Ceylon. The cylindrical body is covered with smooth scales which are somewhat larger on the ventral than on the dorsal surface. The pointed head tapers into the body with no obvious restriction at the neck. With the exception of one genus (*Platyplectrurus*), the barely functional eye is covered with an ocular scale. In most species, the short tail ends with an enlarged shield, which may be covered with tubercles and adorned with spines. The function of the tail appears to be to anchor the animal in its burrow (for example when dealing with prey). Teeth are present on the upper and lower jaws. They feed mainly on invertebrates, particularly earthworms and burrowing insect larvae. Ovoviviparous, producing 3-15 relatively large young. As they are not frequently kept in captivity, little is known of their husbandry. A tall, narrow, glass terrarium is recommended, this being filled with a loose, well-drained substrate (peat and sand mixture).

FAMILY XENOPELTIDAE—SUNBEAM SNAKE

A family containing only a single genus.

FAMILY COLUBRIDAE—TYPICAL SNAKES

This family is one of the remaining five families of advanced snakes in the infraorder Caenophidia. The Colubridae is the largest and most complex family in the suborder Serpentes and contains some 14 subfamilies, 290 genera and about 2000 species found in tropical, subtropical, and temperate regions throughout the world. Needless to say, many aspects of the taxonomy of such a large family are often under revision. The size range is 20-400 cm (8 in—13 ft). Habits and habitats are extremely variable and there are diurnal or nocturnal, terrestrial or arboreal, and aquatic or fossorial forms. Reproduction may be oviparous or ovoviviparous.

Subfamily Xenoderminae

The Xenoderminae contains the most primitive colubrids which show certain superficial resemblances to the wart snakes (Acrochordidae). There are 7 genera and about 12 species native to Southeast Asia, Central and South America. Usually found in humid areas, in swamps or tropical rainforest. The biology of the members of this subfamily is poorly known and there are opportunities for the terrarium keeper to contribute to further research.

Subfamily Sibynophinae—Many-toothed Snakes

With three genera and about 17 species native to Southeast Asia, Madagascar and Central America. Found in rainforest and hill forest. Length from 30-80 cm (12-32 in). The head is not set off distinctly from the neck. A characteristic of the family is the numerous, fragile teeth, 25-56 on each maxillary, which are flattened and dagger-like, an adaptation connected with a diet of hard-scaled lizards (such as skinks).

Subfamily Xenodontinae

This subfamily contains 27 genera and about 193 species, all native to the Americas. The species are small to medium-sized snakes, some with well-developed opisthoglyph fangs (rear-fanged, venomous). The subfamily contains species of highly varying forms and habits. All species in the subfamily appear to be oviparous.

Subfamily Calamarinae

Contains nine genera and some 65 species distributed from India to New Guinea and the Philippines. Found mainly in rainforest and montane rainforest to an altitude of 1800 m (5850 ft). Length from 15-75 cm (6-30 in). Most have the head indistinct from the neck and a pointed or shovel-shaped snout. The rostral shield usually extends forward over the mouth. Scales are smooth and iridescent. The relatively short tail usually ends with a spine. All species are subterranean, living in burrows, only emerging at night or after heavy rain. Food consists mainly of earthworms, other invertebrates, and their larvae. Some species live in termite mounds and feed on the termites and their nymphs. All are best kept in glass terraria suitable for burrowing snakes.

Genera include: *Agrophis* F. Mueller, 1894; *Brachyorrhus* Kuhl, 1826; *Calamaria* Boie, 1827; *Idiopholis* Mocquard, 1892; *Pseudorabdion* Jan, 1862; *Rabdion* (= *Rhabdophidium*) Dumeril, 1853; and *Typhlogeophis* Guenther, 1879.

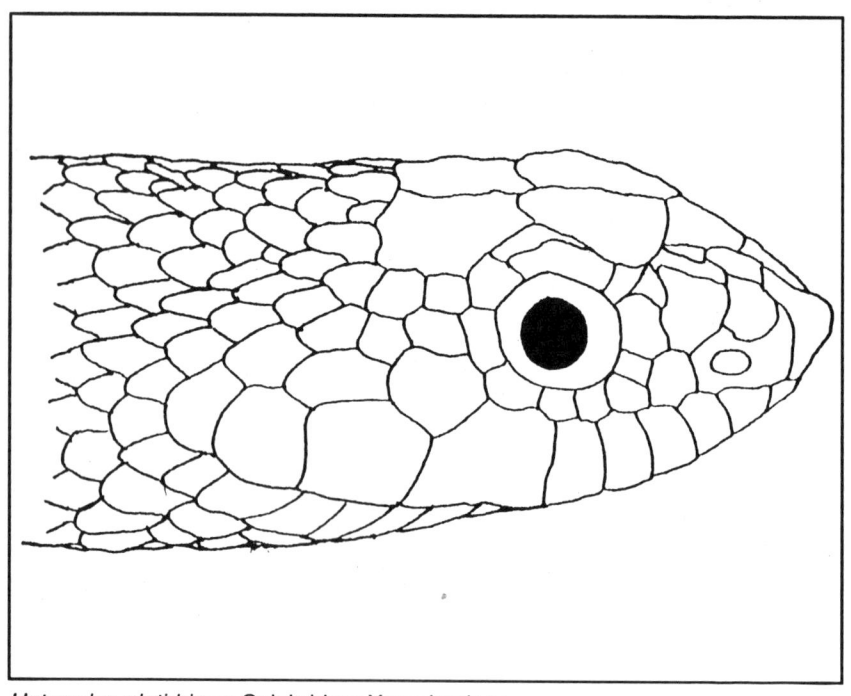

Heterodon platirhinos: Colubridae: Xenodontinae.

Subfamily Colubrinae

With about 50 genera and some 300 species. Most genera occur in Africa, Asia, and the Americas, with just a few in Europe and only a single genus in Australia. Many of the species are relatively long (from 150-300 cm—60-120 in total adult length). They can be be described as typical snakes, most of them being slender and agile. None of them possess venom fangs although there are some really aggressive species in the group. Methods of prey capture include the "grab and swallow," and constriction. Some of the genera are specialized in certain prey (e.g. snail eaters).

Subfamily Dasypeltinae—African Egg-eating Snakes

The subfamily contains only a single genus.

Subfamily Lycodontinae

With about 51 genera and some 325 species found mainly in Africa and Asia, but with a few genera (doubtfully placed here and possibly better placed in Natricinae or Colubrinae) in North and South America. Small to medium snakes. The subfamily status is concerned mainly with the teeth arrangement. Large recurved teeth designed for catching and holding prey are arranged in groups at the front and back of the upper jaw and at the front of the lower jaw, an arrangement which has led to the common name of wolf snakes for many of the species.

Subfamily Natricinae—Water Snakes
There are about 37 genera and some 185 species in the subfamily, and representatives of these are found in most temperate, subtropical, and tropical parts of the world. Although generally referred to as water snakes, many of the species are only semi-aquatic, while some are terrestrial, fossorial, or even semi-arboreal. Lengths range from about 30-250 cm (12-100 in). Typical subfamily characteristics include arrangements of the head and body scales, but there is a remarkable lack of uniformity in areas such as tooth arrangement, some of them being opisthoglyphs (rear-fanged), others being totally non-venomous.

Subfamily Homalopsinae
This subfamily contains 11 genera and some 35 species ranging from India through Indo-China and the Indo-Australian Archipelago to N. Australia. They are very aquatic snakes, found in fresh to brackish water; some species are even seen far out to sea and they have populated estuaries and mangrove swamps on many islands. All are opisthoglyphic, with well-formed, grooved venom fangs in the rear of the upper jaw. Although the

Cemophora coccinea: Colubridae: Colubrinae.

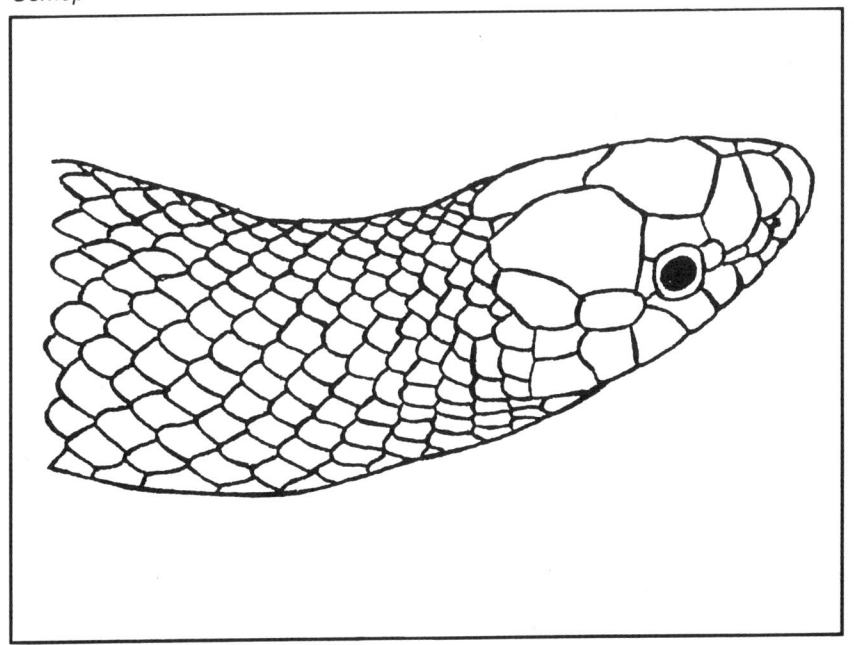

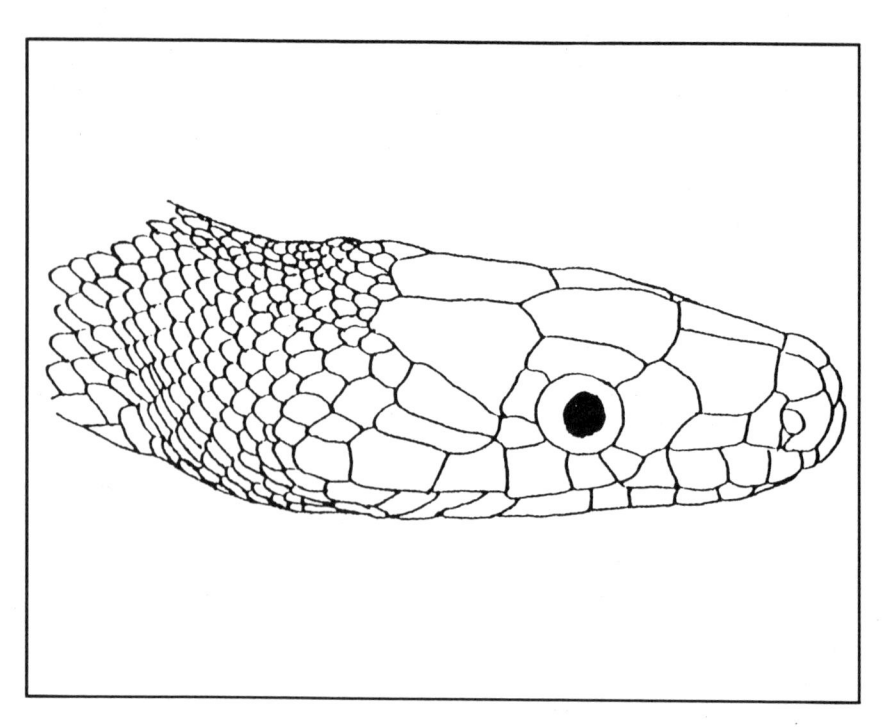

Above: *Elaphe moellendorffi:* Colubridae: Colubrinae. **Below:** *Nerodia erythrogaster flavigaster:* Colubridae: Natricinae.

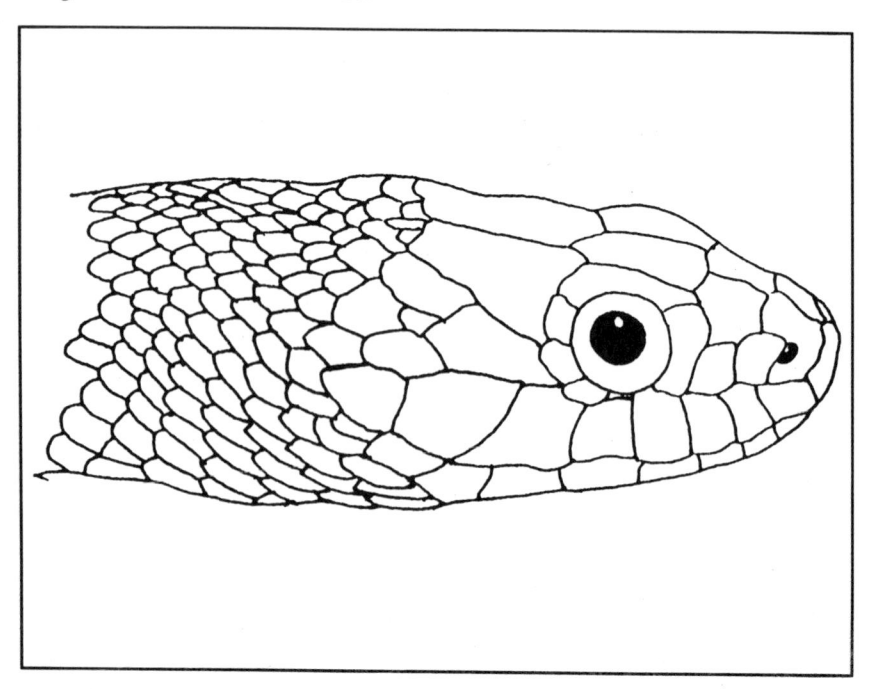

venom is not proven to be dangerous to humans, these species should be treated with the respect any venomous snake deserves. Eyes and valved nostrils are directed upward, a typical characteristic of aquatic animals.

Subfamily Boiginae

Containing about 73 genera and some 330 species, the majority of which are native to Africa and tropical America. A few genera are found in N. America, Europe, and Asia, while Australia has just a single genus. All are opisthoglyphic with grooved venom fangs at the rear of the upper jaw, these varying in size depending on the species. The toxicity of the venom to humans is varied but fatal bites have been recorded from the genera *Dispholidus*, *Thelotornis*, and *Boiga*. All species in the subfamily should be treated with respect, remembering that the reaction to venom varies from person to person and that antivenom is available only for the highly dangerous Boomslang (*Dispholidus typus*). The species have colonized varying habitats, and details of their captive care will depend on the climatic zones and microclimates to which they

Ahaetulla prasina: Colubridae: Boiginae.

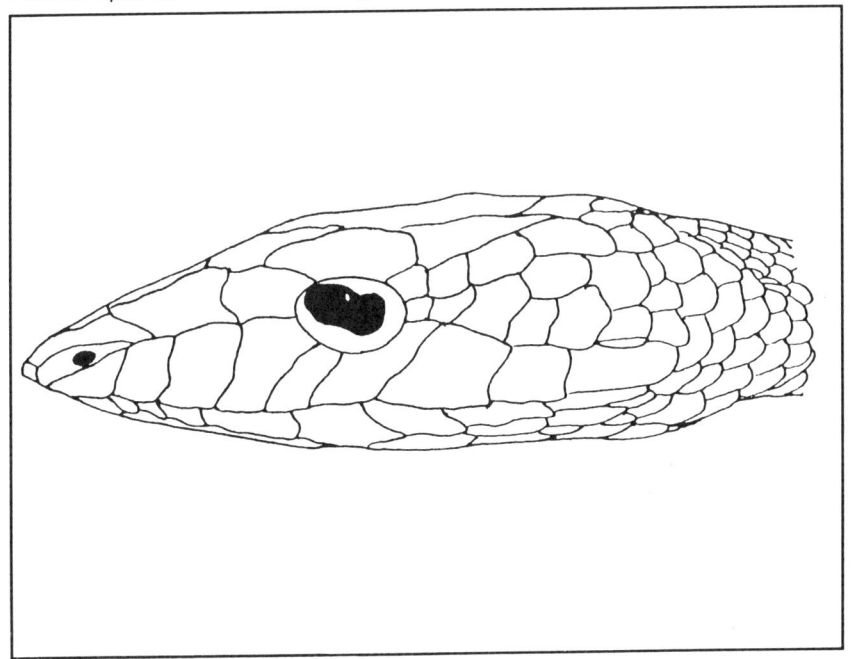

are native. With a few isolated exceptions, all species in the subfamily are oviparous.

As with all subfamilies in the Colubridae, there is considerable controversy about whether or not this is a natural group. Many authorities consider the Boiginae to be an artificial grouping representing varied lineages of otherwise unrelated snakes that have independently developed the rear-fang condition. Such authorities doubt that the Old and New World rear-fangs are related, but instead represent parallel evolution from unrelated ancestors. The Boiginae (like the other subfamilies) should be considered a convenient group but not necessarily a "real" subfamily.

Subfamily Dipsadinae—American Snail-eating Snakes

This subfamily contains 3 genera and about 48 species native to Central and South America in rainforest and montane forest. Lengths range from 25-90 cm (10-36 in). Highly specialized, some species feed exclusively on snails, which are extracted from their shells by the special long teeth at the front of the lower jaw. They require a heated rainforest terrarium. Local garden snails will usually be accepted as food.

Subfamily Pareinae—Asian Snail-eating Snakes

With two genera and about 15 species native to S.E. Asia, in monsoon, rain, or montane rainforest. Similar in many respects to Dipsadinae (parallel evolution). Length to 90 cm (36 in). Head distinct from laterally compressed body. Large eyes with vertical pupils. Long anterior teeth in lower jaw. Feed almost exclusively on snails. Care similar to that described for Dipsadinae.

Subfamily Elachistodontinae— Indian Egg-eating Snakes

A subfamily containing only a single genus and a single species.

Subfamily Aparallactinae—Mole Vipers

Containing about ten genera with 45 or so species native to Africa and the Middle East, the taxonomy of this subfamily has caused many problems, and some species have previously been contained in the subfamily Boiginae or even in the families Elapidae or Viperidae. Recent revision places all genera in the Aparallactinae, with the possible exception of *Atractaspis*, which may warrant its own subfamily (Atractaspinae). Many species in the subfamily have relatively long, opisthoglyph or solenoglyph fangs set forward in the jaw and supplied with large venom glands. Like those of viperids, the fangs of some species are hinged and lie against the jaw until ready to be erected as necessary. Although the affects of their venom on **man**

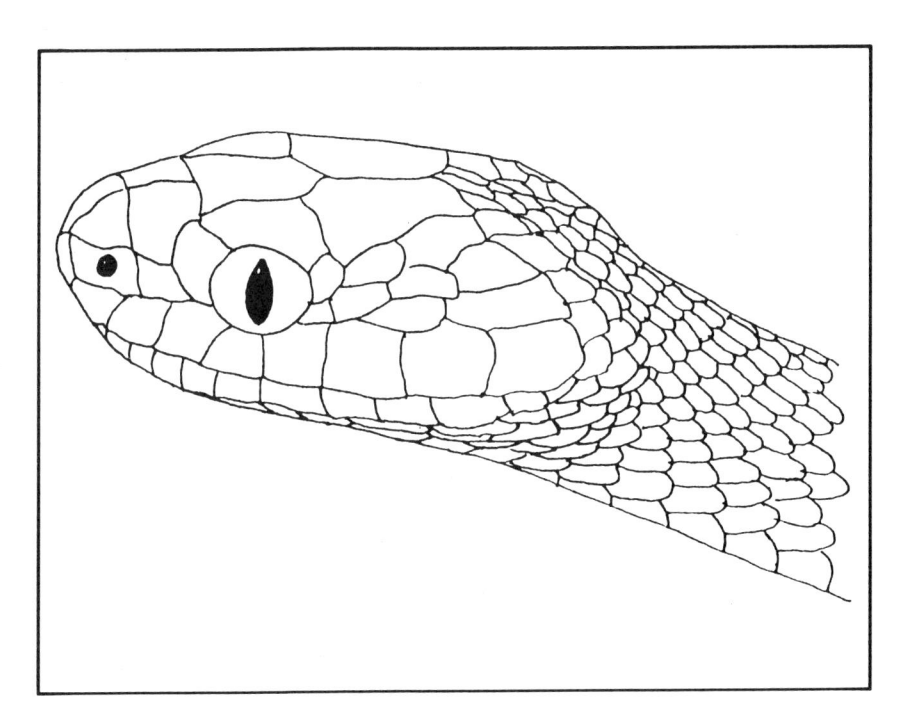

Above: *Boiga ocellata:* Colubridae: Boiginae. **Below:** *Psammophylax tritaeniatus:* Colubridae: Boiginae.

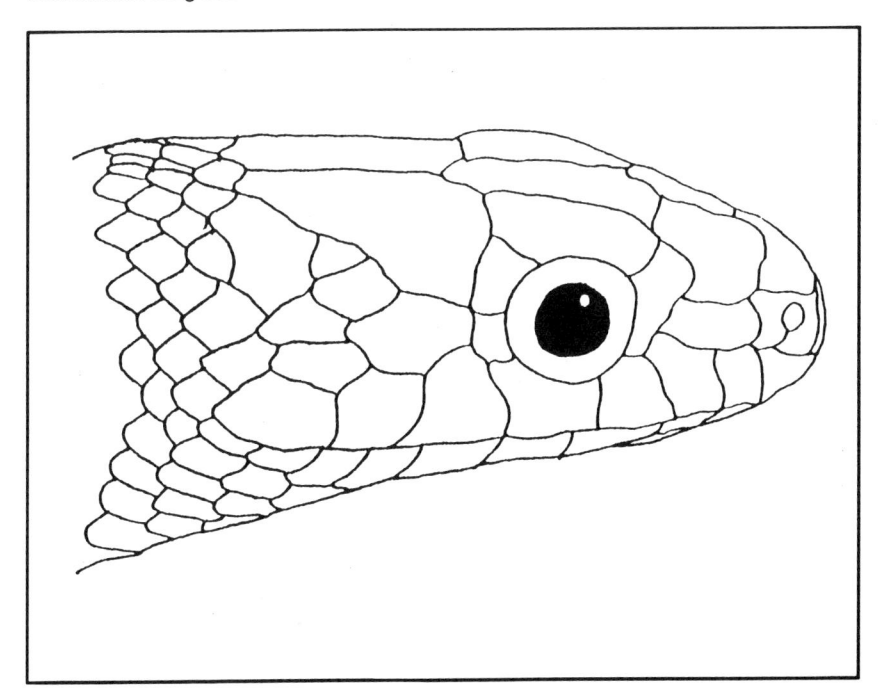

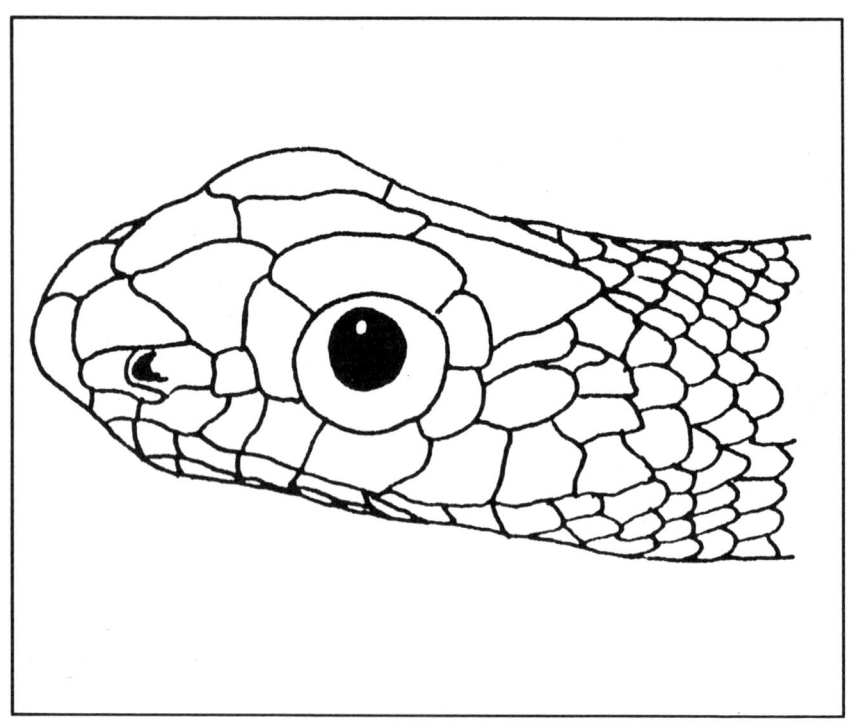

Malpolon moilensis: Colubridae: Boiginae.

are not fully researched, all species in the group should be regarded as venomous and dangerous. Bites from *Atractaspis*, for example, can cause intense pain and swelling. Most species in the subfamily are burrowers and should be kept in a terrarium with a deep, loose substrate.

FAMILY ELAPIDAE—COBRAS, MAMBAS, KRAITS, CORAL SNAKES, ETC.

With about 50 genera and some 200 species of venomous snakes distributed throughout the tropics and subtropics and only absent from most of the temperate zones and the colder parts of the Northern Hemisphere. The family has its headquarters in Australia, where elapids are the ruling group of snakes. Thus Australia has the distinction of being the only country where venomous snake species outnumber nonvenomous ones.

There have been some recent suggestions that the majority of Australian elapids should be assigned to the Hydrophiidae, but most workers still prefer to leave them in the Elapidae. Lengths of most species are in the range of 30-100 cm

(12-40 in); notable exceptions are the King Cobra, *Ophiophagus hannah*, with a record length of 5.6 m (18.2 ft) and the Taipan, *Oxyuranus scutellatus*, with a record length of 4 m (13 ft). Most of the species are slender and colubrid-like in form, but there are a few exceptions (for example, *Acanthophis*). The upper side of the head is typically covered with large scales. All species possess a pair of fixed proteroglyph fangs at the front of the upper jaw. These are relatively short when compared with the fangs of the Viperidae and the Crotalidae. The venom of most elapids is mainly neurotoxic, with only minimal hemotoxic effects in most cases. There are, however, exceptions. Some species are highly dangerous to humans, and all species in the group must be handled with the greatest of care. They are not recommended for the home terrarium-keeper and are best left to the attentions of the zoo-keeper or professional herpetologist.

Elapids have found their way into a variety of habitats. There are those

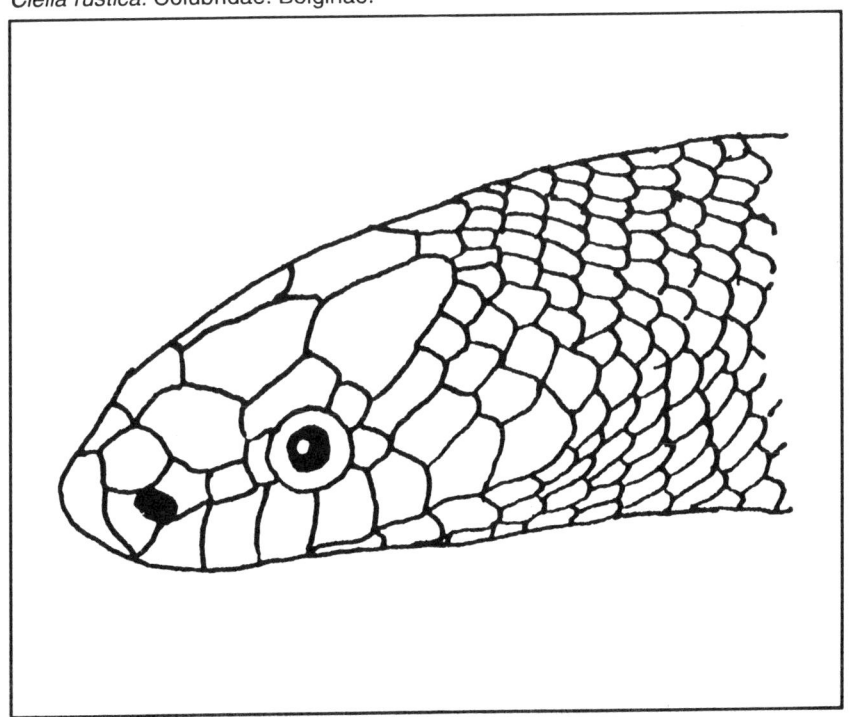

Clelia rustica: Colubridae: Boiginae.

that are primarily terrestrial (including fossorial types), those that are primarily arboreal, and a few that are semi-aquatic. They may be oviparous or ovoviviparous.

FAMILY HYDROPHIIDAE—SEA SNAKES

Comprising two subfamilies, 16 genera, and about 56 species native to the Pacific and Indian Oceans and adjacent tropical and subtropical seas, occasionally venturing accidentally into temperate waters. One or two species have possibly become endemic to freshwater lakes. Length to 275 cm (110 in). Most are strongly adapted to a marine existence, with laterally compressed bodies and rudder-like tails to aid in swimming. Most have a tendency to reduce or lose the broad ventral scales typical of most other snake families, these being replaced by scales of a size similar to the dorsals. As with most aquatic snakes, the eyes are relatively small. The nostrils are valved and kept closed when the reptiles are submerged. Sea snakes possess, in the floor of the mouth surrounding the tongue sheath, a gland that is adapted to removing excessive salt from the bloodstream, the salt being expelled when the tongue is extended. The tooth structure is similar to that of the Elapidae: short proteroglyphic venom fangs at the front of the upper jaw followed by varying numbers of smaller, backwardly directed teeth on the maxillas and dentaries and usually on the palatines and pterygoids. Sea snakes feed largely on fish, especially eels. With a single exception (*Laticauda*) all genera are ovoviviparous, giving birth to live young in the water, and thus have no need to come onto land to nest. Some live near the shore and make excursions out to sea. *Laticauda* must come onto land in order to lay its eggs.

Hydrophiidae is considered a doubtful family by many herpetologists. *Laticauda* is very similar to *Bungarus* and allies (Elapidae), and probably most workers would now remove it from the true sea snakes. The Hydrophiinae proper could be treated as a full family (Hydrophiidae) or a subfamily of an expanded Elapidae. One school of thought, not generally accepted currently, combines the Australian elapids with the hydrophiids to produce a greatly enlarged Hydrophiidae and a very small and uniform family Elapidae. The systematics of these groups is sure to change in the future.

Sea snakes have rarely been kept in captivity, thus their husbandry is poorly documented, though in recent times some public aquaria have exhibited them **in huge seawater tanks and**

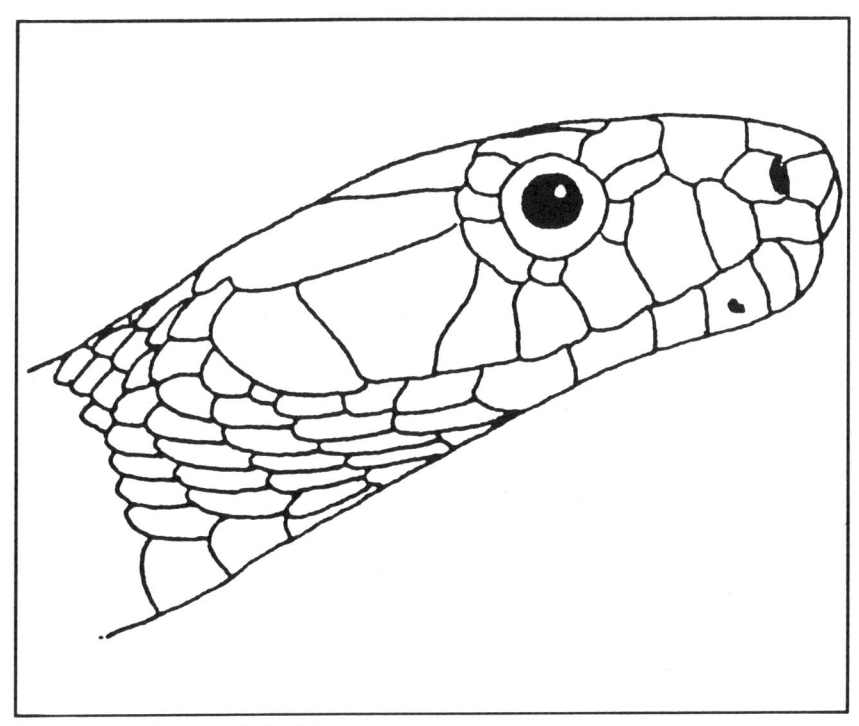

Dendroaspis jamesoni: Elapidae.

marine biologists are making progress in understanding their husbandry. In the wild, many species migrate considerable distances, probably coinciding with their breeding habits, and it can be assumed that they require a large amount of space in order to behave normally. Additionally, they require the same relatively complicated conditions demanded in aquaria by marine fish. Most sea snakes do not transport well. Once removed from their watery environment they are subject to internal injuries caused by collapse of the body and they also cannot respire adequately. The venom of most species is considered highly toxic to humans although most species seem reluctant to bite out of the water. Nevertheless, they should **not** be kept by anyone other than a qualified professional; zookeepers, scientists, etc. The professional herpetologist who also has an interest in marine aquaria has an amost unlimited opportunity to discover new and exciting factors concerning the biology of sea snakes.

Subfamily Laticaudinae—
This subfamily contains only a single genus which many authorities now

believe to be a genus of Elapidae and not a true sea snake. Others believe the subfamily to be a family in its own right, Laticaudidae, sea kraits.

Subfamily—Hydrophiinae

With 15 genera and some 51 species found mainly in coastal areas and estuaries (as described for family), but some may be found far out to sea. One genus (*Hydrophis*) may be found in freshwater lakes. Unlike Laticaudinae, which have normal nostrils, the nostril of Hydrophiinae are set high back on the snout to increase efficiency of respiration. The short venom fangs are back in the upper jaw on the shortened maxilla and may be preceded by short palatine teeth. All show the tendency to lose the broad ventral scales, these being replaced more or less with small scales similar to the dorsals. The musculature of the body is strongly reduced and most species find it difficult or impossible to move or respire on land. All are ovoviviparous, giving birth to 2-6 large, well-developed young in the water.

FAMILY VIPERIDAE—TYPICAL OR OLD WORLD VIPERS

With three subfamilies, 11 genera, and about 49 species native to Europe, Asia, and Africa in tropical to temperate climates. Most authorities include the pit vipers here as a subfamily Crotalinae. One species (*Vipera berus*) reaches within the Arctic Circle in Europe. Many species are adopted to desert conditions, but others are found in savannah to tropical rain forest. Many species are active at night but may spend much of the day basking. They possess a pair of relatively long solenoglyphic venom fangs that rest along the roof of the mouth and are brought forward by rotation of the maxilla as the reptile strikes at prey or in defense. In most cases prey is released after envenomation. The prey animal dies in a few minutes and the snake then follows its scent trail and devours it. All species in the family are highly venomous and dangerous to man, mostly producing hematoxic/cytotoxic venom, and strict safety precautions are required when such species are kept in captivity.

Subfamily Azemiophinae—Fea Vipers

This subfamily contains a single genus.

Subfamily Viperinae—True Vipers

With nine genera and some 46 species found in Europe, Africa, and Asia. Length 30–200 cm (12-80 in) Characterized in most cases by broad, triangular head distinct from robust body by relatively narrow neck. Head covered generally with small scales instead of large plates. Dorsal scales partially to strongly keeled.

Many have a characteristic zig-zag pattern, others are uniformly colored or have an amazing multicolored pattern. Most species more or less ovoviviparous.

Subfamily Causinae—Night Adders

Contains only a single genus.

FAMILY CROTALIDAE—PIT VIPERS

With at least seven genera and some 130 species native mainly to the Americas, but with some representatives in S. and S.E. Asia. Some authorities include this family as the subfamily Crotalinae within the Viperidae. Members of the family are found in a range of climates from cool to warm temperate, subtropical, and tropical. Length 60-350 cm (24-140 in). Large, often triangular head distinct from robust body by relatively narrow neck. The more primitive types (*Agkistrodon*, *Sistrurus*) possess large, symmetrical head shields, while the majority have a head covered with small scales. The main characteristic of the family is the pit, a thermoreceptor

Vipera berus: Viperidae: Viperinae.

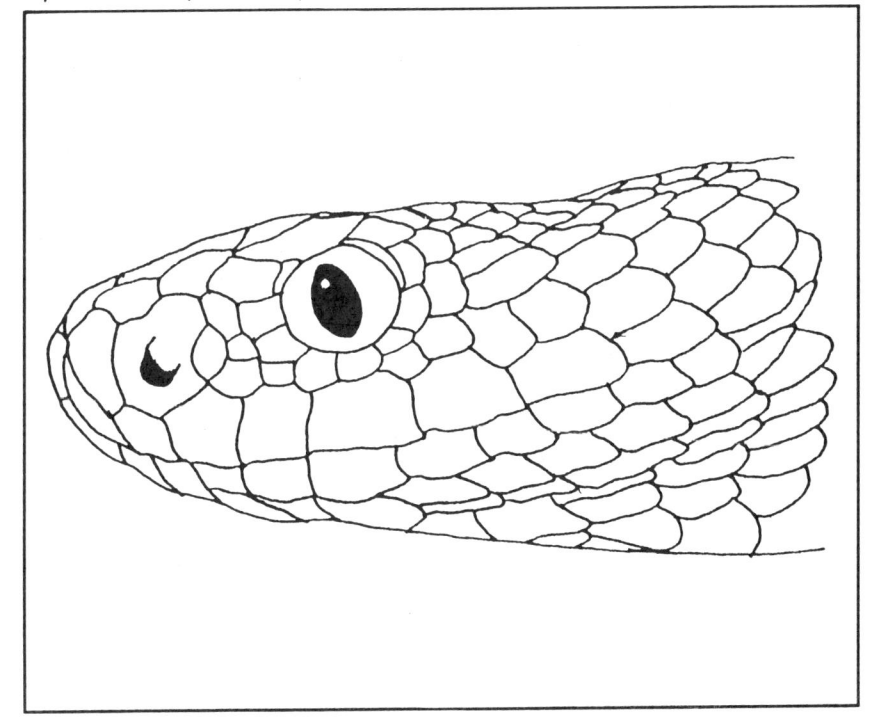

organ situated on each side of the head between the nostril and the eye. This organ is similar in function but more efficient than the pits on the boid lip. It contains an infra-red heat-sensitive diaphragm that enables the reptile to seek prey and orient in the dark.

There are large solenoglyphic fangs in the front of the upper jaw. Most of the species are highly venomous, and all require very secure housing. Other factors pertaining to housing are dependent on the natural habitat of the individual species. Most are ovoviviparous, but there are exceptions.

Atheris squamiger: Viperidae: Viperinae.

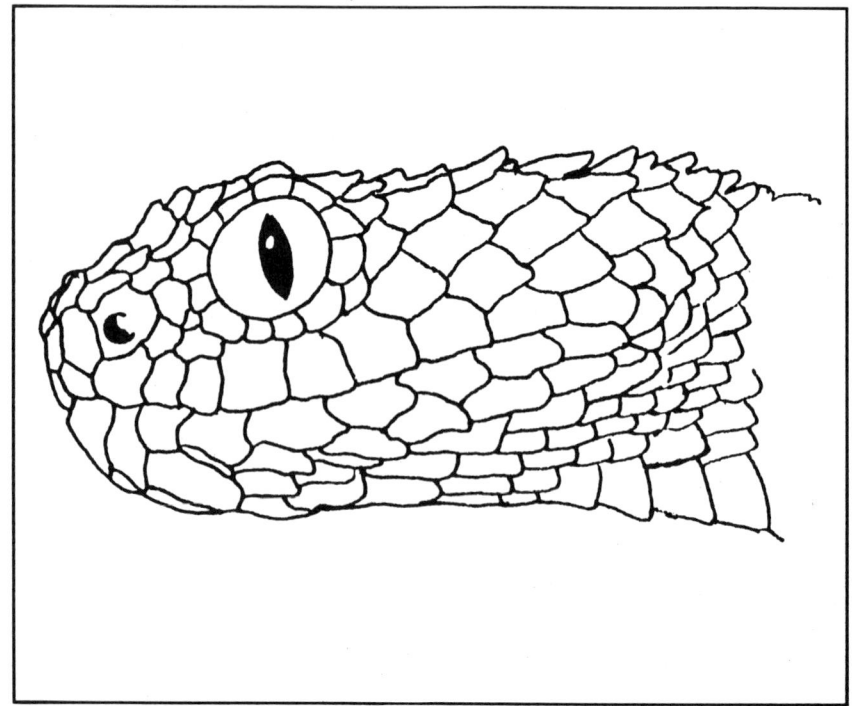

Listing of Snake Genera
arranged by family and subfamily

FAMILY TYPHLOPIDAE
Ramphotyphlops
Rhinotyphlops
Typhlops

FAMILY ANOMALEPIDAE
Anomalepis
Helminthophis
Liotyphlops
Typhlophis

FAMILY LEPTOTYPHLOPIDAE
Leptotyphlops
Rhinoleptus

FAMILY ANILIIDAE
Anilius
Anomochilus
Cylindrophis

FAMILY BOIDAE
Subfamily Loxoceminae
Loxocemus

Subfamily Pythoninae
Aspidites
Calabaria
Chondropython
Liasis
Morelia
Python

Subfamily Boinae
Acrantophis
Boa
Corallus
Epicrates
Eunectes
Exiliboa
Lichanura
Sanzinia
Tropidophis
Ungaliophis

Subfamily Bolyeriinae
Bolyeria
Casarea

Subfamily Erycinae
Candoia
Charina
Eryx

FAMILY UROPELTIDAE
Brachyophidium
Melanophidium
Platyplecturus
Plecturus
Pseudotyphlops
Rhinophis
Teretrurus
Uropeltis

FAMILY XENOPELTIDAE
Xenopeltis

FAMILY COLUBRIDAE
Subfamily Xenoderminae
Achalinus
Cercaspis
Fimbrios
Nothopsis
Stoliczkia
Xenodermus
Xenopholis

Subfamily Sibynophinae
Liophidium
Scaphiodontophis
Sibynophis

Subfamily Xenodontinae
Alsophis
Antillophis
Arrhyton
Conophis
Cyclagras
Darlingtonia
Ditaxodon
Dromicus
Heterodon
Hypsirhynchus
Ialtris
Leimadophis
Lioheterophis
Liophis
Lygophis
Lystrophis
Paroxyrhopus
Philodryas
Platyinion
Rhadinaea
Rhadinella
Sordellina
Synophis
Tretanorhinus
Umbrivaga
Uromacer
Uromacerina
Xenodon

Subfamily Calamarinae
Agrophis
Brachyorrhos
Calamaria
Calamorhabdium
Etheridgeum
Idiopholis
Pseudorabdion

Rabdion
Typhlogeophis

Subfamily Colubrinae
Arizona
Cemophora
Chilomeniscus
Chionactis
Chironius
Coluber
Conopsis
Contia
Coronella
Dendrelaphis
Dendrophidion
Drymarchon
Drymobius
Drymoluber
Duberria
Eirenis
Elaphe
Eurypholis
Gastropyxis
Gonyophis
Hapsidophrys
Hydrablabes
Iguanognathus
Lampropeltis
Leptodrymus
Leptophis
Liopeltis
Lytorhynchus
Masticophis
Mastigodryas
Meizodon
Opheodrys
Philothamnus
Phyllorhynchus
Pituophis
Prosymna
Pseudaspis
Pseudoficimia
Pseustes
Ptyas
Rhinocheilus
Rhynchophis
Salvadora
Scaphiophis
Simophis
Spalerosophis
Spilotes
Stilosoma
Thrasops
Zaocys

Subfamily Dasypeltinae
Dasypeltis

Subfamily Lycodontinae
Adelphicos
Anoplohydrus
Aspidura
Atractus
Blythia
Boaedon
Bothrolycus
Bothrophthalmus
Chamaelycus
Compsophis
Cryptolycus
Cyclocorus
Dendrolycus
Dinodon
Dromicodryas
Dryocalamus
Elapoidis
Farancia
Geagras
Geophis
Glypholycus
Gonionotophis
Haplocercus
Haplonodon
Hormonotus
Lamprophis
Lepturophis
Leioheterodon
Liopholidophis
Lycodon
Lycodonomorphus
Lycognathophis
Lycophidion
Mehelya
Micropisthodon
Ninia
Oligodon
Opisthotropis
Oreocalamus
Pararhadinaea
Plagiopholis
Pseudoxyrhopus
Rhabdops
Rhynchocalamus
Stegonotus
Tetralepis
Trachischium
Tropidodipsas
Xylophis

Subfamily Natricinae
Adelophis
Afronatrix
Amastridium
Amphiesma
Atretium
Balanophis
Carphophis
Chersodromus
Diadophis
Diaphorolepis
Gonyosoma
Grayia
Helicops
Hydrops
Limnophis
Macropisthodon
Macropophis
Natriciteres
Natrix
Nerodia
Pxyrhabdium
Paraptychophis
Pararhabdophis
Pliocercus
Pseudoeryx
Pseudoxenodon
Ptychophis
Regina
Rhabdophis
Seminatrix
Sinonatrix
Storeria
Thamnophis
Trimetopon

Tropidoclonion
Virginia
Xenelaphis
Xenochrophis

Subfamily Homalopsinae
Bitia
Cerberus
Enhydris
Erpeton
Fordonia
Homalopsis
Hurria
Myron

Subfamily Boiginae
Ahaetulla
Alluaudina
Amplorhinus
Apostolepis
Boiga
Chamaetortus
Choristocalamus
Chrysopelea
Clelia
Coniophanes
Crotaphopeltis
Dipsadoboa
Dispholidus
Drepanoides
Dromophis
Dryophiops
Elapomojus
Elapomorphus
Enulius
Erythrolamprus
Ficimia
Geodipsas
Gomesophis
Gyalopion
Hemirhagerrhis
Hologerrhum
Hydrodynastes
Hypoptophis
Hypsiglena
Imantodes
Ithycyphus
Langaha

Leptodeira
Lycodryas
Macroprotodon
Madagascarophis
Malpolon
Manolepis
Mimophis
Opisthoplus
Oxybelis
Oxyrhopus
Parapostolepis
Phimophis
Procinura
Psammodynastes
Psammophis
Psammophylax
Pseudablabes
Pseudoboa
Pseudoleptodeira
Pseudotomodon
Pythonodipsas
Rhachidelus
Rhamphiophis
Rhinobothryum
Scolecophis
Siphlophis
Sonora
Stenorrhina
Symphimus
Sympholis
Tachymenis
Tantilla
Tantillita
Telescopus
Thamnodynastes
Thelotornis
Toluca
Tomodon
Trimorphodon
Tripanurgos
Xenocalamus

Subfamily Dipsadinae
Dipsas
Sibon
Sibynomorphus

Subfamily Pareinae

Aplopeltura
Pareas

Subfamily Elachistodontinae
Elachistodon

Subfamily Aparallactinae
Amblyodipsas
Aparallactus
Atractaspis
Calamelaps
Chilorhinophis
Homoroselaps
Macrelaps
Micrelaps
Miodon
Polemon

FAMILY ELAPIDAE
Acanthophis
Aspidelaps
Aspidomorphus
Austrelaps
Boulengerina
Bungarus
Cacophis
Calliophis
Cryptophis
Demansia
Dendroaspis
Denisonia
Drysdalia
Echiopsis
Elapognathus
Elapsoidea
Furina
Glyphodon
Hemachatus
Hemiaspis
Hoplocephalus
Leptomicrurus
Maticora
Micruroides
Micrurus
Naja
Neelaps
Notechis

Ogmodon
Ophiophagus
Oxyuranus
Parademansia
Paranaja
Parapistocalamus
Pseudechis
Pseudohaje
Pseudonaja
Rhinoplocephalus
Salomonelaps
Simoselaps
Suta
Toxicocalamus
Tropidechis
Unechis
Vermicella
Walterinnesia

FAMILY HYDROPHIIDAE
Subfamily Laticaudinae
Laticauda

Subfamily Hydrophiinae
Acalyptophis
Aipysurus
Disteira
Emydocephalus
Enhydrina
Ephalophis
Hydrelaps
Hydrophis
Lapemis
Parahydrophis
Pelamis
Thalassophis

FAMILY VIPERIDAE
Subfamily Azemiophinae
Azemiops

Subfamily Viperinae
Adenorhinos

Atheris
Bitis
Cerastes
Daboia
Echis
Eristicophis
Pseudocerastes
Vipera

Subfamily Causinae
Causus

FAMILY CROTALIDAE
Agkistrodon
Bothrops
Crotalus
Hypnale
Lachesis
Sistrurus
Trimeresurus

Photographers and Illustrators

Both John Coborn and TFH would like to thank and acknowledge the following people for their superb photographic work, without which this book simply would not have been possible:

C. Banks	R. Everhart	R. L. Holland
R. D. Bartlett	A. v. d. Nieuwenhuizen	R. A. Winstel
P. Freed	R. S. Simmons	F. Achaval
G. Pisani	M. Freiberg	A. I. Grasso
A. Kerstitch	H. Piacentini	H. Nicolay
Dr. S. A. Minton	J. Iverson	A. Norman
R. Kayne	R. Holland	L. Edmonds
J. Merli	W. Wuster	L. Trutnau
W. Tomey	J. Visser	N. Gray
K. H. Switak	J. Wines	B. Carlson
B. Kahl	J. Coborn	D. Reed
W. P. Mara	W. B. Allen, Jr.	W. Deas
P. J. Stafford	D. Beckwith	R. Steene
G. Dingerkus	G. Marcuse	H. Schultz
The London Zoo	S. B. Reichling	M. Dee
K. T. Nemuras	J. T. Kellnhauser	J. Dommers
B. Christie	R. E. Kuntz	S. C. and H. Miller
The Chester Zoo	M. J. Cox	P. Vargas
K. Lucas	H. Hansen	W. E. Burgess
S. Kochetov	J. K. Langhammer	Suzanne L. & Joseph
R. T. Zappalorti	J. Gee	T. Collins

The two black and white illustrations on page 5 were taken from Rev. Wood's *Animate Creation*. In addition, the beautiful head and dorsal drawings in the *Lampropeltis* pages were done by John R. Quinn, and both the line drawings in the systematic section and the symbols used throughout the pictorial section were done by W. P. Mara.
If there is anyone who has not been given proper credit where it was due, we apologize for the oversight.

Key to the Symbols

In the following pictorial section you will notice a line of tiny symbols below each photograph. These have been included to give you a few extra bits of information concerning each snake depicted. The topics these symbols correspond to are: 1) what food item(s) the pictured snake accepts most often; 2) whether the snake is an egglayer or a livebearer; 3) and whether or not the snake is venomous.

A more detailed description of each symbol is included here:

🐀 MAMMAL-EATER

This means the snake in question is a small mammal-eater, not simply a mouse-eater. It could also take rats, shrews, voles, or any other item it could secure.

🐟 FISH- OR CRUSTACEAN-EATER

Most aquatic snakes will accept fish most eagerly, but some species (*Regina septemvittata*, for example) feed entirely on crayfish. If you find yourself in possession of one of the snakes that boasts this symbol, you may have to do some feeding experiments to find out *exactly* what it will eat.

🐦 BIRD-EATER

Many tree-dwelling species, or at the very least those snakes that will go into trees looking for food, may include both adult and nestling birds in their diet.

🦎 REPTILE-EATER

Many snakes feed on reptiles, and not just lizards but other snakes as well. Keepers usually dread keeping these types of snakes because acquiring food on a regular basis can be very difficult if not impossible.

🐸 AMPHIBIAN-EATER

Salamanders (including newts and amphiumas) and frogs and toads are all part of the diet of certain snake species. These items can sometimes be very difficult for a keeper to supply.

🕷 INSECT- AND/OR SMALL INVERTEBRATE-EATER

Tiny ground-dwelling snakes are notorious for preying on crickets, earthworms, slugs, spiders, and the like. Sometimes you can find these items right in your own backyard.

● EGG-EATER

There are a few snake species that count eggs as a large part of their diet. Bird eggs are the primary focus, but reptile eggs account for some of it as well.

⚭ EGG-LAYER

The term "oviparous" refers to those snakes that lay clutches of eggs. Snake eggs are most delicate and cannot be fooled with. They have a leathery outer shell and some have a rough, granular texture.

⚯ LIVE-BEARER

The term for a snake that gives birth to living young is "ovoviviparous" (or simply "viviparous"). From a professional breeder's point of view, this method is superior to egglaying in that the embryos have already fully developed and incubation of eggs is unnecessary.

☠ VENOMOUS

This tells you that the snake depicted has been judged "potentially harmful or known to be harmful" through the injection of poison channeled through fangs, but does not necessarily mean the snake has caused human fatalities. Snakes with this symbol should not be handled freely or trusted to any degree.

"X" cm

Finally, you will notice a measurement of length (in centimeters) at the end of each line. This represents the average maximum adult length for the snake depicted, rounded down to the nearest 5 cm. Thus, a snake with an average maximum adult length of 123 cm would be recorded at 120 cm. These lengths are not meant as *exact* references, but will give you some idea of how long the snake in question grows under normal circumstances.

PICTORIAL SECTION

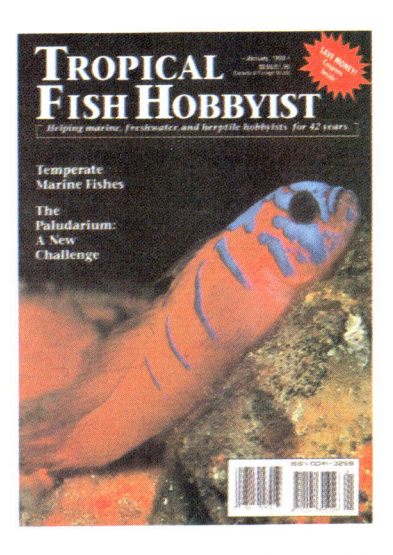

Since 1952, *Tropical Fish Hobbyist* has been the source of accurate, up-to-the-minute, and fascinating information on every facet of the aquarium hobby. Join the more than 50,000 devoted readers worldwide who wouldn't miss a single issue.

Subscribe right now so you don't miss a single copy!

Return To:
Tropical Fish Hobbyist, P.O. Box 427, Neptune, NJ 07753-0427

YES! Please enter my subscription to *Tropical Fish Hobbyist*. Payment for the length I've selected is enclosed. U.S. funds only.

CHECK ❏ 1 year-$30   ❏ 2 years-$55   ❏ 3 years-$75   ❏ 5 years-$120
ONE: 12 ISSUES   24 ISSUES   36 ISSUES   60 ISSUES

(Please allow 4-6 weeks for your subscription to start.) *Prices subject to change without notice*

❏ LIFETIME SUBSCRIPTION (max 30 Years) $495
❏ SAMPLE ISSUE $3.50
❏ GIFT SUBSCRIPTION. Please send a card announcing this gift. I would like the card to read: _____
❏ I don't want to subscribe right now, but I'd like to have one of your FREE catalogs listing books about pets. Please send catalog to:

SHIP TO:
Name _____
Street _____ Apt. No. _____
City _____ State _____ Zip _____

U.S. Funds Only. Canada add $11.00 per year; Foreign add $16.00 per year.
Charge my: ❏ VISA   ❏ MASTER CHARGE   ❏ PAYMENT ENCLOSED

Card Number   Expiration Date

Cardholder's Name (if different from "Ship to":)

Cardholder's Address (if different from "Ship to":)

                                                      Cardholder's Signature

...From T.F.H., the world's largest publisher of bird books, a new bird magazine for birdkeepers all over the world...

CAGED BIRD HOBBYIST
IS FOR EVERYONE
WHO LOVES BIRDS.

CAGED BIRD HOBBYIST
IS PACKED WITH VALUABLE
INFORMATION SHOWING HOW
TO FEED, HOUSE, TRAIN AND CARE
FOR ALL TYPES OF BIRDS.

Subscribe right now so you don't miss a single copy! SM-316

Return to:
CAGED BIRD HOBBYIST, P.O. Box 427, Neptune, NJ 07753-0427

YES! Please enter my subscription to **CAGED BIRD HOBBYIST**. Payment for the number of issues I've selected is enclosed. *U.S. funds only.

CHECK ONE:		
☐	4 Issues	$9.00
☐	12 Issues for the Price of 10	25.00
☐	1 Sample Issue	3.00

☐ Gift Subscription. Please send a card announcing this gift.   PRICES SUBJECT TO CHANGE
I would like the card to read _____

☐ I don't want to subscribe right now, but, I'd like to receive one of your FREE catalogs listing books about pets. Please send the catalog to:

SHIP TO:
Name _____ Phone ( ) _____
Street _____
City _____ State _____ Zip _____

U.S. Funds Only. Canada, add $1.00 per issue; Foreign, add $1.50 per issue.

Charge my: ☐ VISA   ☐ MASTER CHARGE   ☐ PAYMENT ENCLOSED

Card Number _____ Expiration Date _____

Cardholder's Name (if different from "Ship to:") _____

Cardholder's Address (if different from "Ship to:") _____

Please allow 4-6 weeks for your subscription to start.   Cardholder's Signature

Ramphotyphlops nigrescens, 🕷 🥚 40cm

Typhlops schlegeli, 🕷 🥚 70cm

Liotyphlops albirostris, 🕷 🐍 30cm

Leptotyphlops dulcis dissectus, ✳ ♂ 25cm
Leptotyphlops humilis, ✳ ♂ 40cm

Anilius scytale, 🢂 〰 70cm

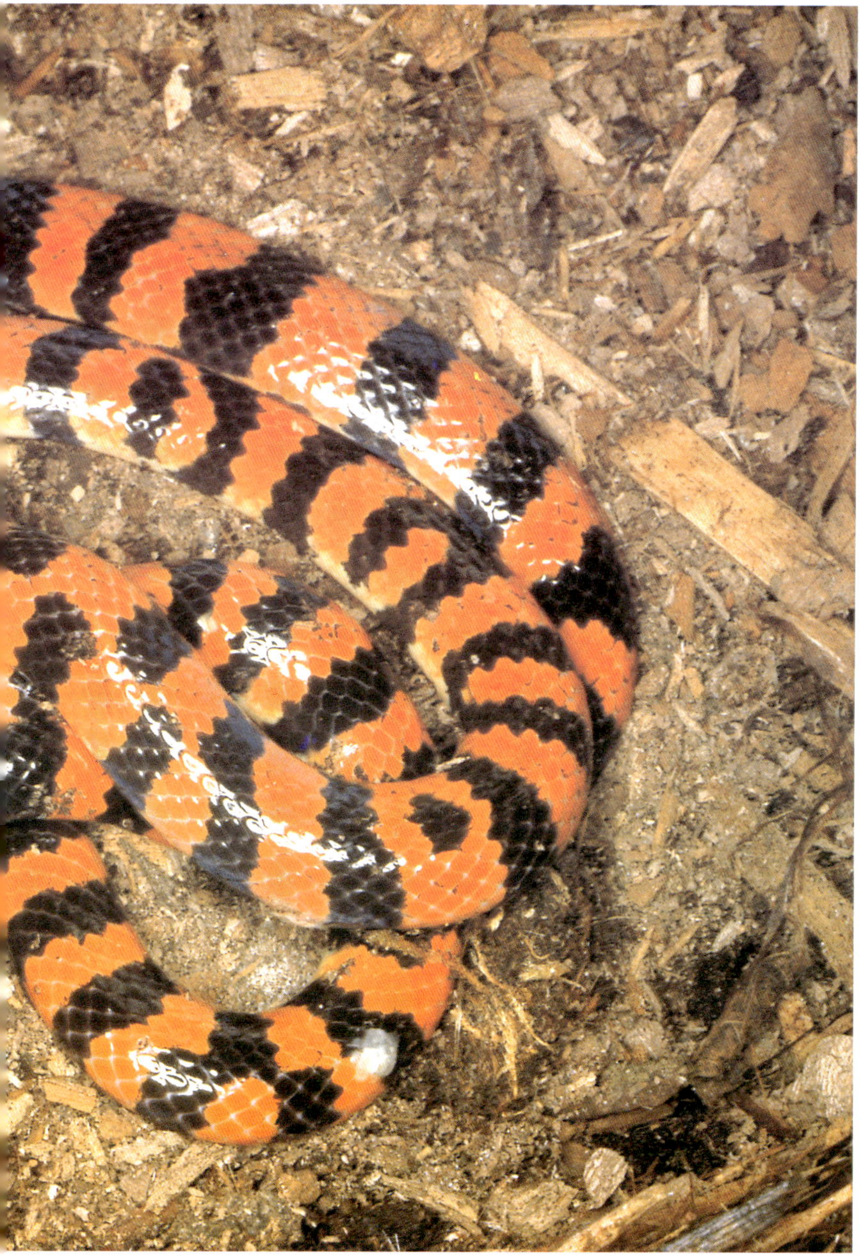

Loxocemus bicolor, 🐍 150cm

Loxocemus bicolor, 🐍 150cm

38

Aspidites ramsayi, 🐭 ☌ 215cm

Aspidites melanocephalus, 🐭 ☌ 180cm

Calabaria reinhardtii, 🐁 ⚭ 100cm

Calabaria reinhardtii, 🐁 ⚭ 100cm

Chondropython viridis, 🐁🐦 ☂ 180cm

Chondropython viridis, 🐀🐦 ☾ 180cm
Chondropython viridis, 🐀🐦 ☾ 180cm

Liasis albertisii, ♂ 215cm

Liasis albertisii, ♂ 215cm

Liasis papuanus, 🐀 ☽ 260cm

Liasis boa, 🐀 ☽ 165cm

44

Morelia spilotes variegata, ♂ ♀ 245cm
Morelia spilotes variegata, ♂ ♀ 245cm

Python boeleni, 🐀 🥚 275cm

Python reticulatus, 🐀 🥚 760cm

Python timorensis, 🐀 ✋ 180cm

Python curtus, 🐀 ✋ 175cm

Python regius, 🐀 ⬤ 155cm

Python curtus brongersmai, 🐀 ⬤ 175cm

Python anchietae, 165cm

Acrantophis madagascariensis, 370cm
Acrantophis dumerili, 300cm

Boa constrictor, 🐀 🦅 320cm

Boa constrictor, 🐀 🦅 320cm

51

Boa constrictor, 320cm

Corallus caninus, 🐦 🌿 300cm

Corallus enydris, 🐦 🌿 200cm

Corallus enydris, 🐦 🐁 200cm

Corallus caninus, 🐢 🐦 🐁 300cm

54

Epicrates subflavus, 🐢 🐍 275cm

Epicrates cenchria, 🐢 🐍 200cm

Epicrates monensis monensis, 🐁 🌀 240cm
Epicrates monensis granti, 🐁 🌀 240cm

Eunectes murinus, 🐟 🐟 〰️ 900cm

Eunectes murinus, 🐟 🐠 🍌 900cm

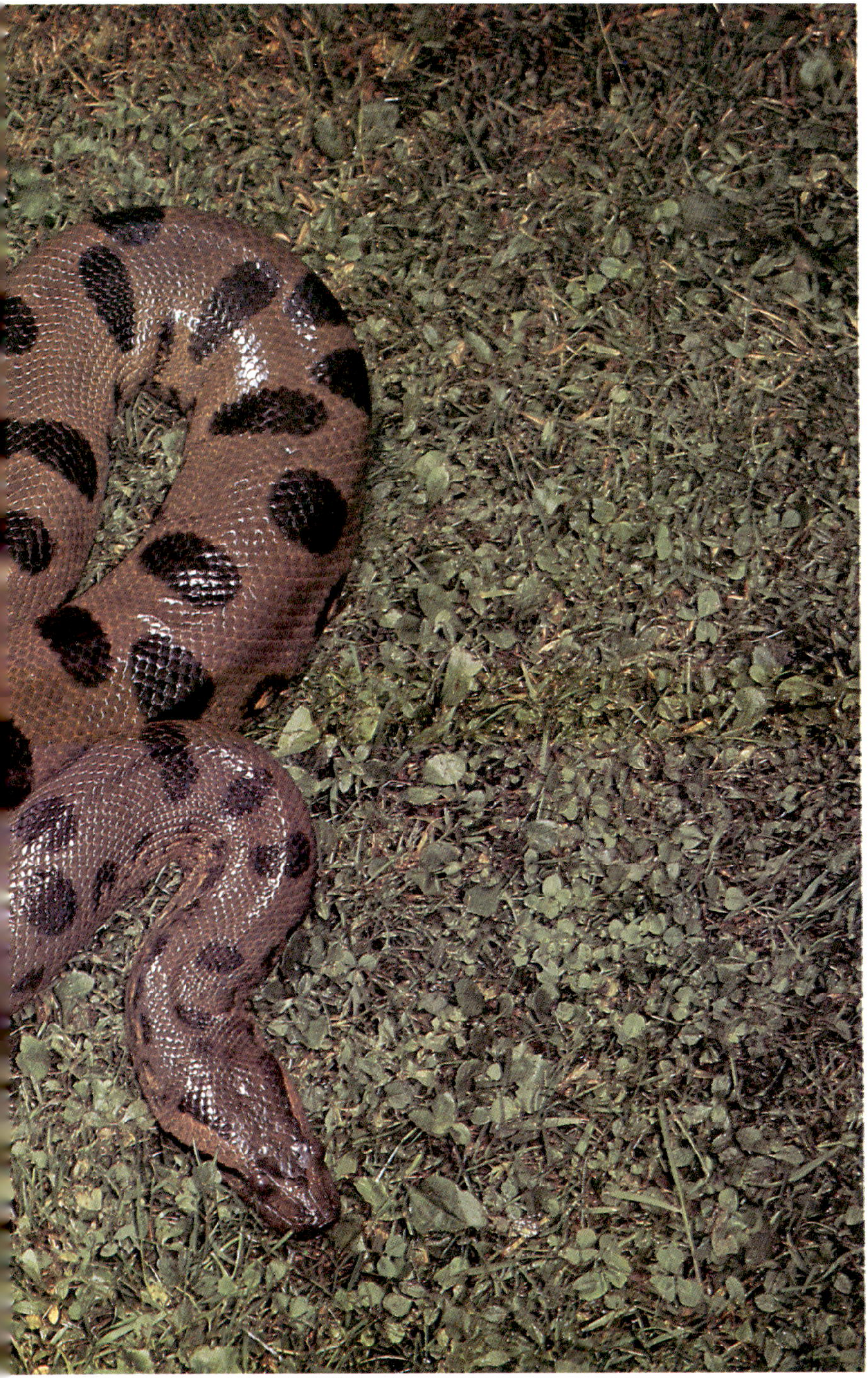

Eunectes notaeus, 🐟 🐟 🐍 400cm

Eunectes murinus, 🐟 🐟 🐍 900cm

Lichanura trivirgata trivirgata, 🐀 🐦 🌀 100cm
Lichanura trivirgata gracia, 🐀 🐦 🌀 110cm

Lichanura trivirgata "myriolepis", 🐢 🐦 🌀 100cm
Lichanura trivirgata trivirgata, 🐢 🐦 🌀 100cm

Sanzinia madagascariensis, 🐢 🐦 🌙 250cm   *Sanzinia madagascariensis,* 🐢 🐦 🌙 250cm

Tropidophis canus curtus, 🦎 🐸 🥚 90cm
Tropidophis melanurus melanurus, 🦎 🐸 🥚 90cm

64

Ungaliophis continentalis, 75cm

Ungaliophis continentalis, 75cm

Candoia aspera, 🐁 🐸 〰 100cm
　　　　　　Candoia carinata paulsoni, 🐁 🐸 〰 90cm

Candoia aspera, 🐭 🐸 ≈ 100cm

Candoia aspera, 🐭 🐸 ≈ 100cm

Charina bottae, 🐁 🐍 80cm

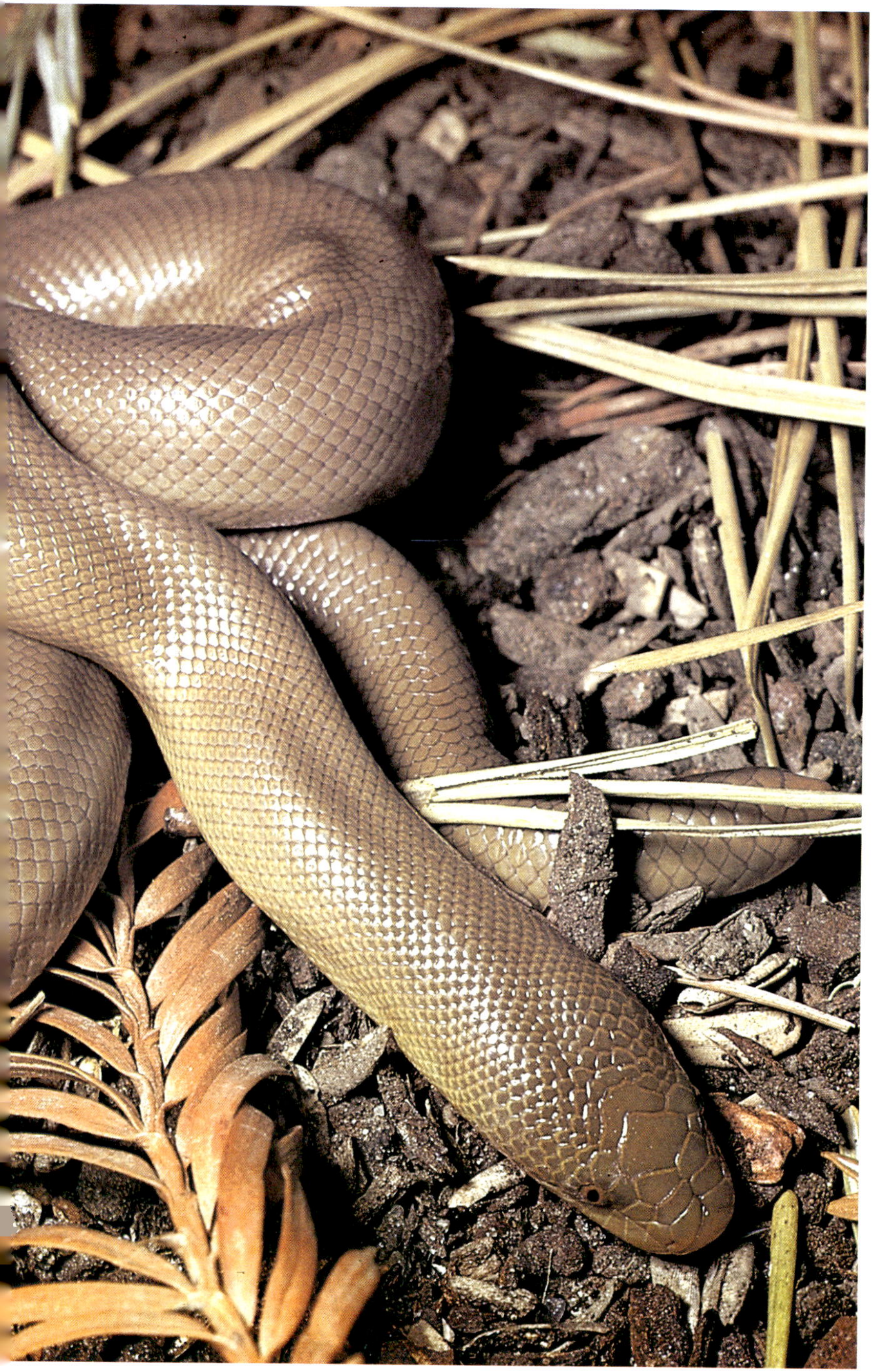

Charina bottae, 🐀 🐁 80cm

Charina bottae, 🐀 🐁 80cm

Eryx tataricus, 🦎 🐍 95cm

Eryx johni, 🐀 🐍 100cm

71

Eryx colubrinus loveridgei, 🐢 🦎 🌙 70cm
Eryx jaculus turcicus, 🐢 🦎 🌙 60cm

Xenopeltis unicolor, 🦎 🐸 🥚 90cm

Xenopeltis unicolor, 🦎 🐸 🥚 90cm

Alsophis vudi picticeps, 🐢 🦎 ⌀ 105cm

Alsophis vudi picticeps, 🐢 🦎 ⌀ 105cm

Conopsis lineatus, ✱ ☽ 35cm

Conopsis lineatus, ✱ ☽ 35cm

Heterodon nasicus, 🐸♂ 80cm

Heterodon nasicus, 🐸♂ 80cm

Heterodon nasicus kennerlyi, 🐢 ♂ 80cm

Heterodon simus, 🐸 ♂ 50cm

77

Heterodon platirhinos, 🐸 ⌒ 85cm

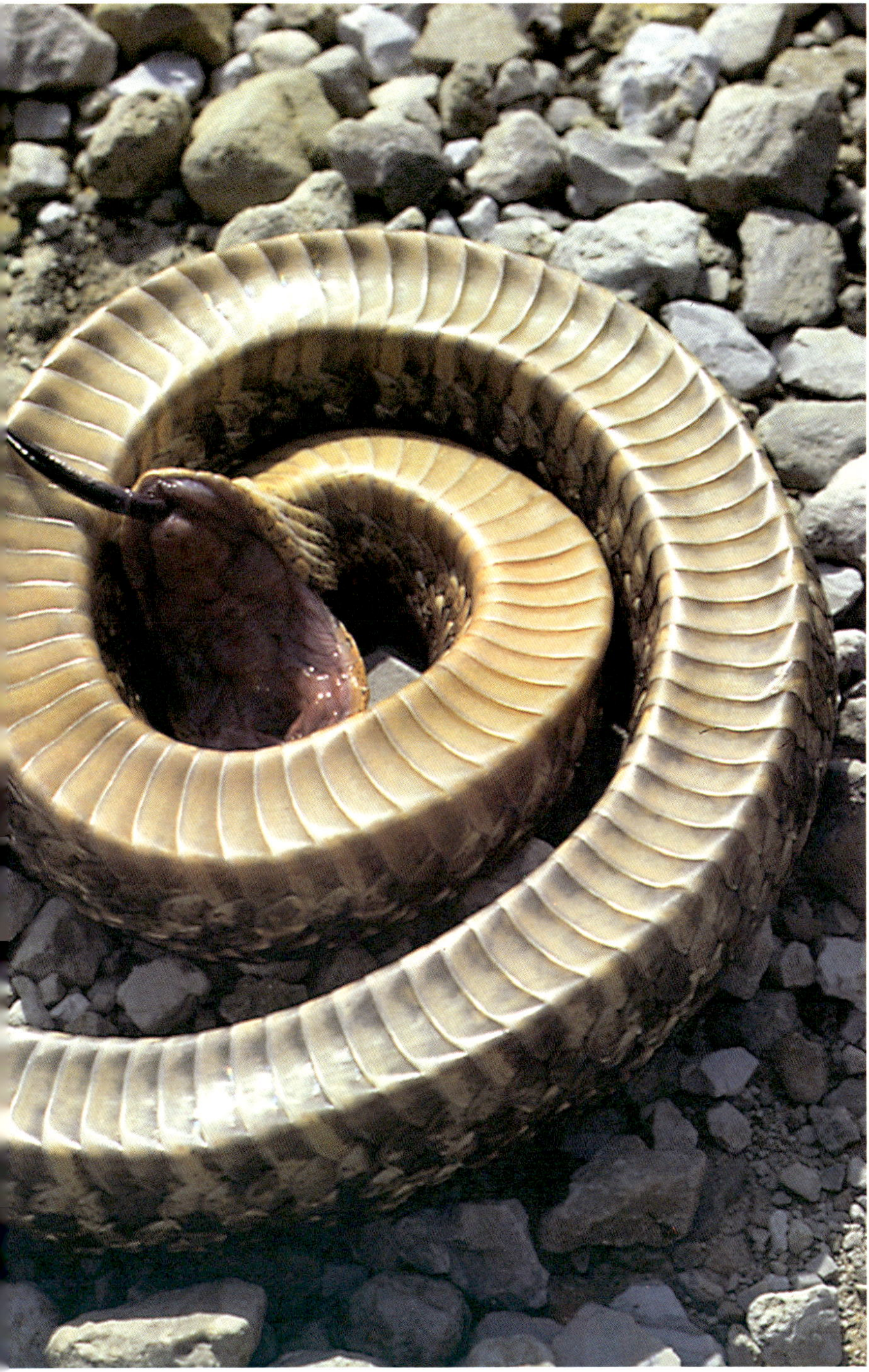

Heterodon platirhinos, 🐸 🥚 85cm

Heterodon platirhinos, 🐸 🥚 85cm

Liophis sp., 🐸 ☾ 70cm

Liophis reginae, 🐸 ☾ 70cm

81

Liophis anomala, 🐸 ♂ 70cm

Liophis epinephalus, 🐸 ♂ 75cm

Lystrophis semicinctus, 🐸 🥚 55cm

Lystrophis dorbignyi, 🐸 🥚 55cm

Lystrophis dorbignyi, 🐸 🥚 55cm

Lystrophis semicinctus, 🐸 🥚 55cm

Philodryas baroni, 🦎 🐸 🥚 120cm
Philodryas psammophideus, 🦎 🐸 🥚 120cm

Rhadinaea brevirostris, 🦎 🐸 ⌁ 50cm

Rhadinaea flavilata, 🦎 🐸 ⌁ 50cm

Uromacer dorsalis, 🦎 🐸 ☂ 160cm   *Uromacer dorsalis,* 🦎 🐸 ☂ 160cm

Xenodon merremi, 🐢 🐸 ⬔ 130cm
　　　　　　　Xenodon rhabdocephalus, 🐢 🐸 ⬔ 130cm

Arizona elegans, 🐁 🐸 🥚 90cm
　　　　　　　　Arizona elegans eburnata, 🐁 🐸 🥚 90cm

89

Arizona elegans philipi, 🐁 🐸 🥚 90cm
Cemophora coccinea coccinea, ⬤ 🥚 50cm

90

Cemophora coccinea, ● ♂ 50cm
Cemophora coccinea copei, ● ♂ 50cm

Cemophora coccinea, ● 🜚 50cm

Chilomeniscus cinctus, ✶ 🜨 25cm

Chilomeniscus cinctus, ✶ 🜨 25cm

Chilomeniscus cinctus, ✷ ◡ 25cm

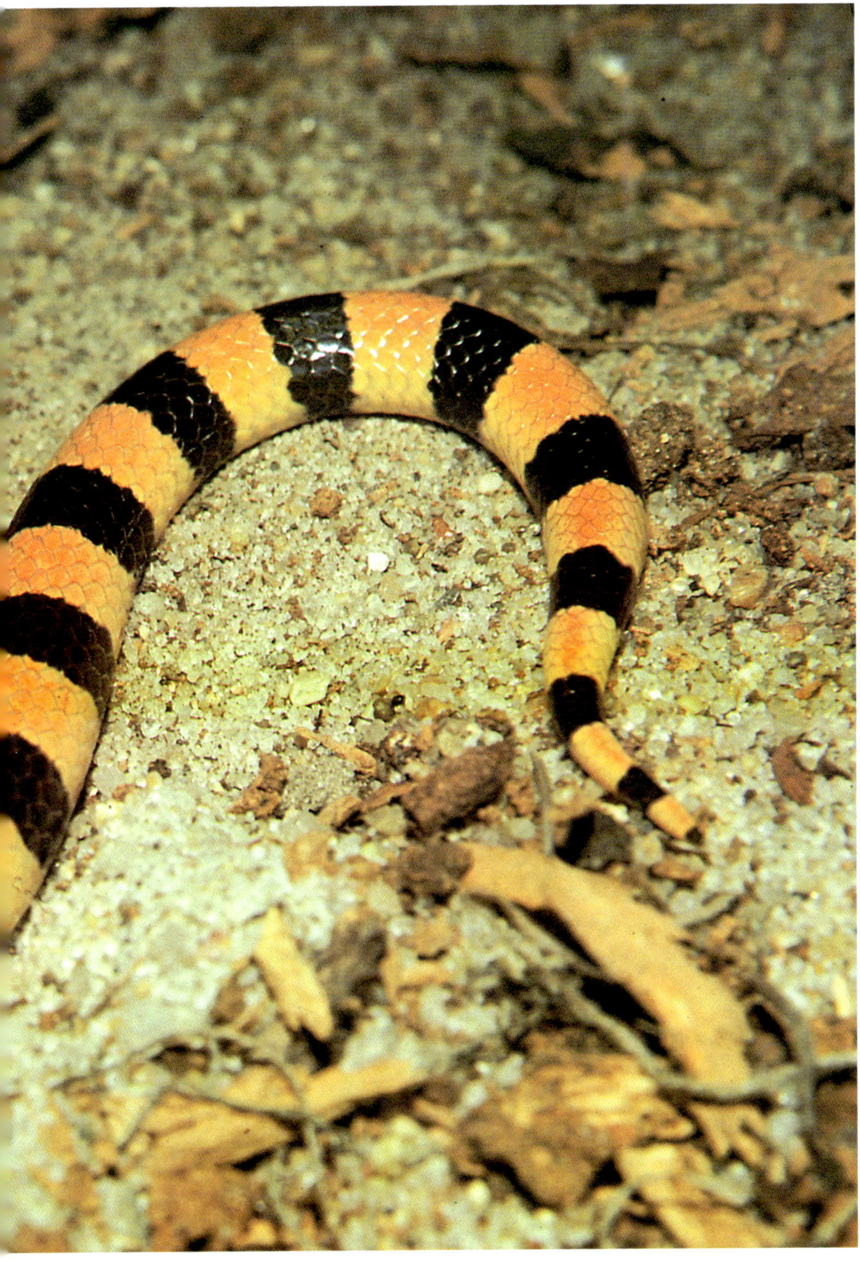

Chionactis palarostris, 🕷 ☾ 35cm

Chionactis occipitalis, 🕷 ☾ 35cm

Chionactis occipitalis klauberi, ✻ ☽ 35cm
Chionactis palarostris, ✻ ☽ 35cm

Coluber ravergieri, 120cm

Coluber hippocrepis hippocrepis, 150cm

Coluber constrictor flaviventris, 🐍 🦎 ✋ 120cm
Coluber constrictor flaviventris, 🐍 🦎 ✋ 120cm

Coluber ravergieri, 120cm
Coluber constrictor latrunculus, 120cm

100

Coluber najadum dahli, 100cm

Coluber ravergieri, 120cm

Coluber karelini mintonorum, 🐭 🐦 ⌥ 100cm
Coluber jugularis, 🐭 🐦 ⌥ 175cm

Coluber constrictor constrictor, 🐍🦎☾ 150cm
Coluber constrictor mormon, 🐍🦎☾ 175cm

Coluber constrictor priapus, 🐍 🦎 ✋ 150cm

Coluber constrictor constrictor, 🐍 🦎 ✋ 150cm

Contia tenuis, 🕷 ☾ 45cm

Coronella austriaca, 🦎 🦫 🐍 70cm

105

Coronella girondica, 🐭 🦎 🥚 70cm
Coronella austriaca, 🐭 🦎 🥚 70cm

Dendrelaphis punctulatus, 🦎 🐸 ☂ 100cm

Dendrelaphis pictus, 🦎 🐸 ☂ 100cm

Dendrelaphis calligaster, 100cm

Drymarchon corais melanurus, 190cm
Drymarchon corais erebennus, 190cm

Drymarchon corais corais, 🐢 ☾ 180cm
　　　　　　　　　Drymarchon corais couperi, 🐢 ☾ 210cm

Drymarchon corais couperi, 🐁 ☾ 210cm
Drymarchon corais melanurus, 🐁 ☾ 190cm

Drymobius margaritiferus, 🐸 ↶ 100cm

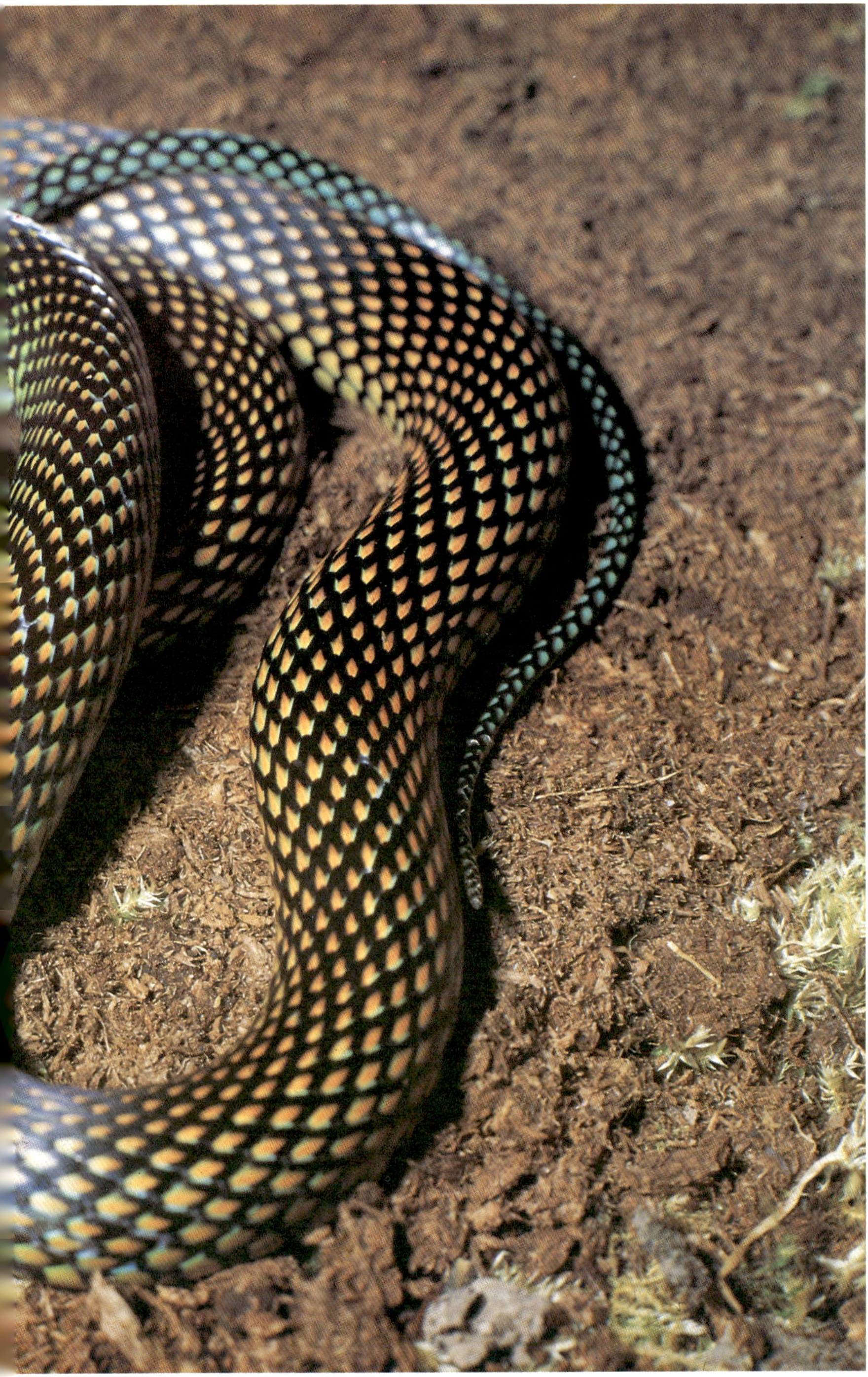

Drymobius margaritiferus, 🐸 ☾ 100cm

Duberria lutrix, 🦎 🕷 ☾ 40cm

Eirenis rothi, 🦎 🦗 🐌 45cm

Eirenis collaris, 🦎 🦗 🐌 45cm

Elaphe [Bogertophis] subocularis, 🐁 ♂ 165cm
Elaphe [Senticolis] triaspis, 🐁 ♂ 150cm

Elaphe obsoleta obsoleta, 🐢 ☾ 170cm

Elaphe guttata guttata, 🐢 ☾ 110cm

Elaphe obsoleta lindheimerii x quadrivittata, 🐁 ☽ 170cm
Elaphe obsoleta lindheimerii, 🐁 ☽ 170cm

Elaphe obsoleta lindheimerii, 🐁 ☌ 170cm
Elaphe guttata guttata, 🐁 ☌ 110cm

Elaphe vulpina vulpina, 🐢 ⚥ 120cm

Elaphe guttata emoryi, 🐢 ⚥ 120cm

Elaphe [Senticolis] triaspis mutabilis, 🐀 ☽ 150cm
Elaphe taeniura ridleyi, 🐀 ☽ 150cm

Elaphe radiata, 🐢 ♂ 150cm

Elaphe [Bogertophis] subocularis, 🐢 ♂ 165cm

Elaphe obsoleta quadrivittata, 🐢 🐦 ☾ 170cm
Elaphe obsoleta "williamsi", 🐢 ☾ 170cm

Elaphe schrenckii, 🐭 🐦 🥚 180 mm

Elaphe scalaris, 🐭 🐦 🥚 120cm

Elaphe rufodorsata, 🐟 🐸 ☾ 75cm
Elaphe [Bogertophis] rosaliae, 🐁 ☾ 140cm

125

Elaphe obsoleta spiloides, 🐀 ☂ 170cm
Elaphe obsoleta quadrivittata, 🐀 🐦 ☂ 170cm

Elaphe obsoleta lindheimerii, 🐢 ♂ 170cm
Elaphe moellendorffi, 🐢 ♂ 200cm

Elaphe obsoleta obsoleta, 🐭 ☾ 170cm

Elaphe moellendorffi, 🐭 ☾ 200cm

Elaphe mandarina, 🐀 🐦 🥚 160cm

Elaphe guttata guttata, 🐀 🥚 110cm

Elaphe guttata guttata, 🐭 ☾ 110cm   Elaphe guttata guttata, 🐭 ☾ 110cm

Elaphe guttata guttata, 110cm

Elaphe guttata guttata, 🐢 ♂ 110cm

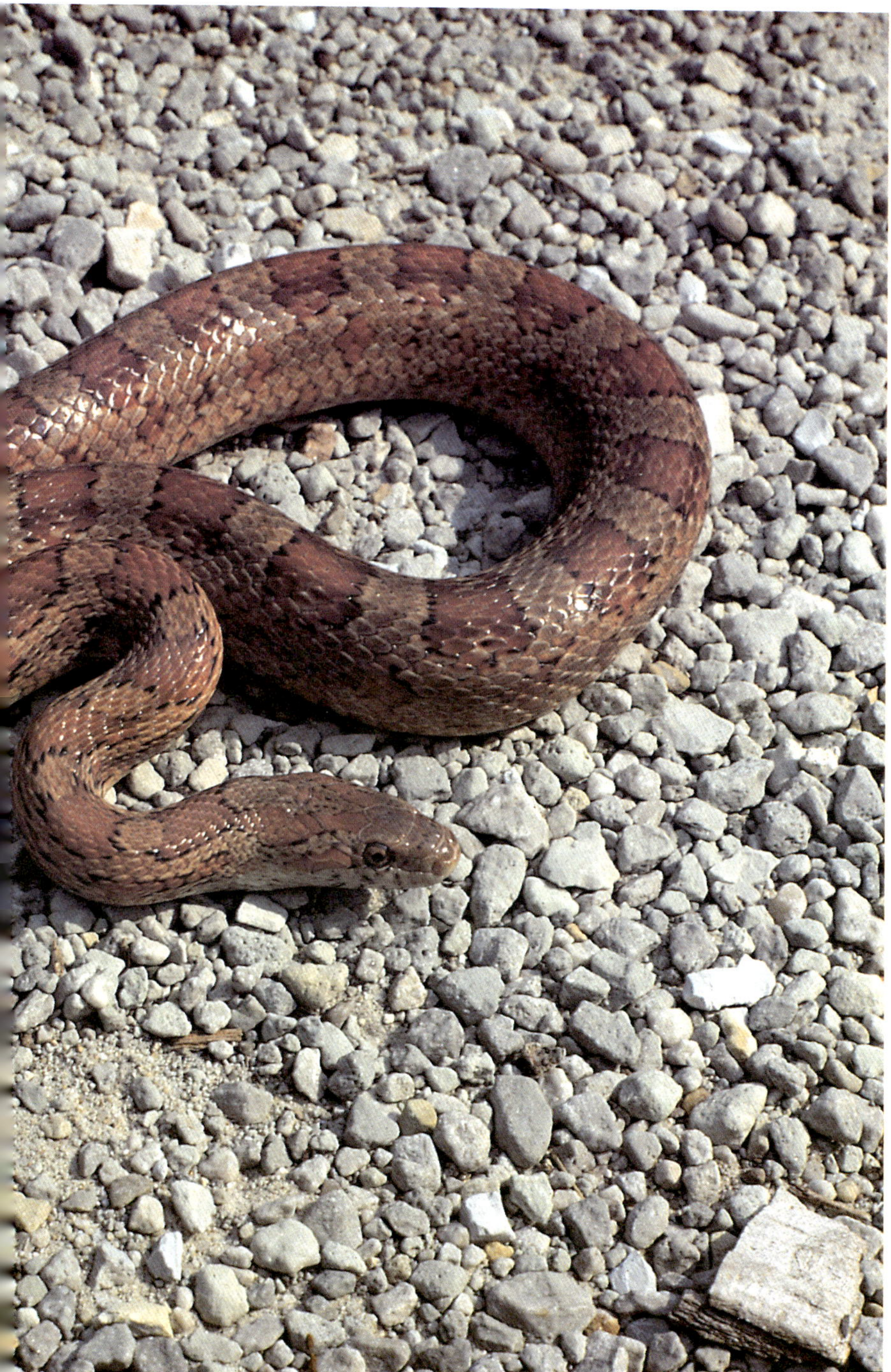

Elaphe helena, 🐭🦎 🌀 95cm

Elaphe carinata, 🦎 🌀 150cm

134

Elaphe [Senticolis] triaspis, 🐢 ☾ 150cm

Elaphe taeniura yunnanensis, 🐢 ☾ 150cm

Elaphe carinata, 150cm

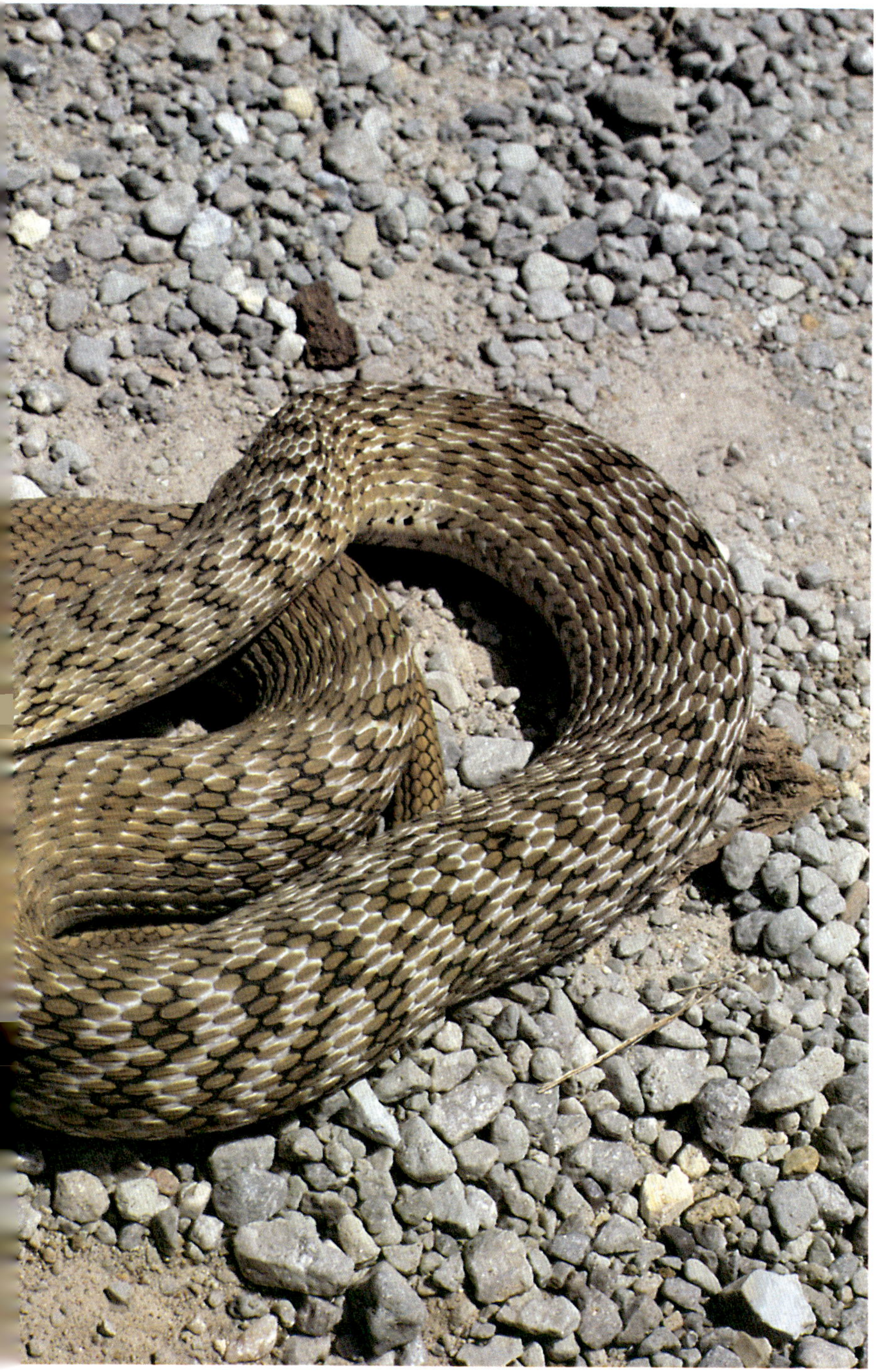

Hapsidophrys lineata, 🐸 ⚥ 110cm

Lampropeltis alterna, 🐁 🦎 ⚥ 120cm

Lampropeltis alterna, 🐀 🦎 ☾ 120cm
Lampropeltis alterna, 🐀 🦎 ☾ 120cm

Lampropeltis alterna, 🐢 🦎 ☾ 120cm

Lampropeltis alterna, 🐊 🦎 ↻ 120cm

Lampropeltis getula goini, 150cm
Lampropeltis getula californiae, 150cm

142

Lampropeltis getula californiae, 🐁 🦎 ☾ 150cm
Lampropeltis getula californiae, 🐁 🦎 ☾ 150cm

143

Lampropeltis getula niger, 110cm

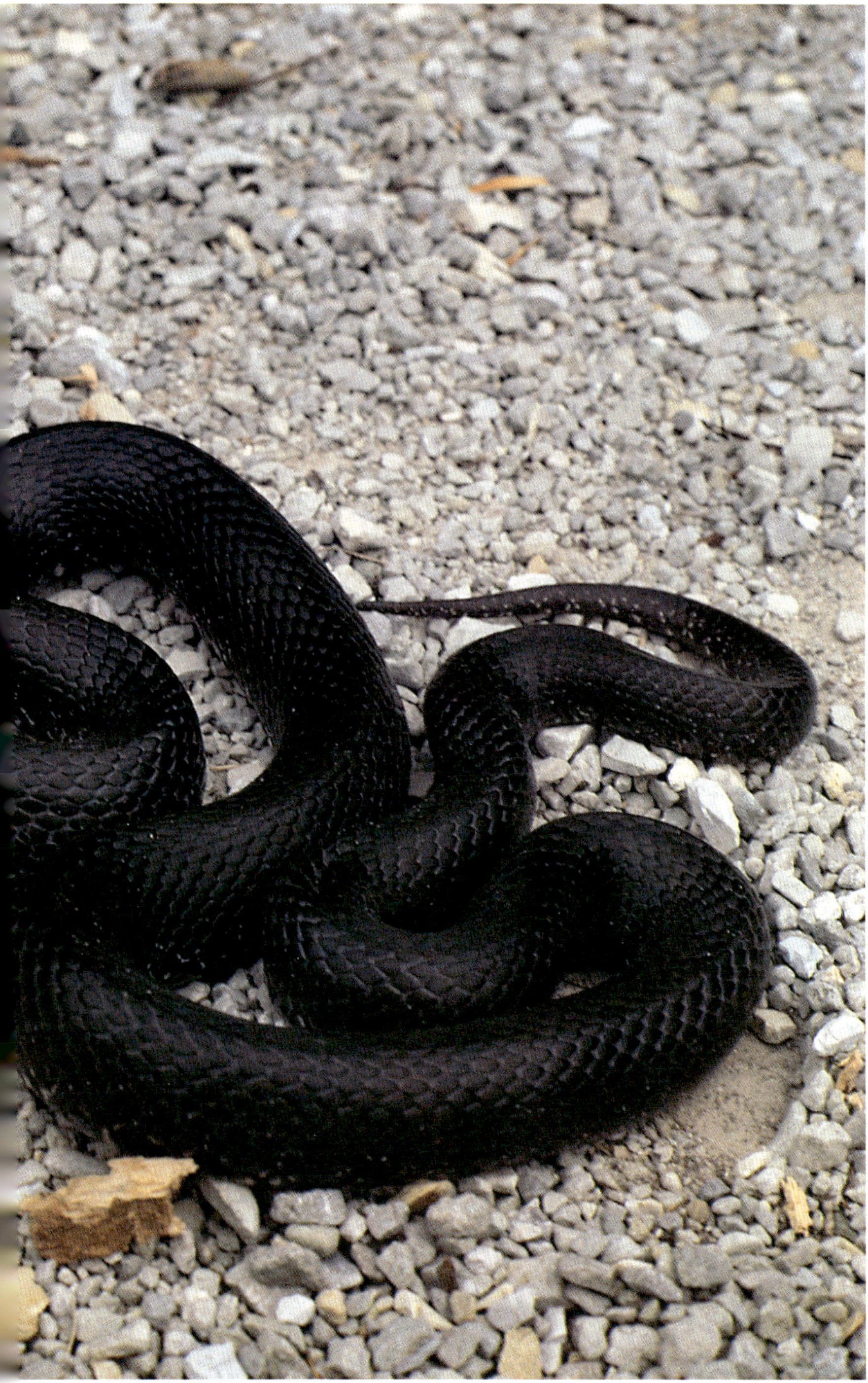

Lampropeltis getula floridana, 150cm
Lampropeltis getula californiae, 150cm

146

Lampropeltis mexicana "greeri", 85cm

Lampropeltis mexicana "thayeri", 🐭 🦎 ♂ 85cm

Lampropeltis mexicana "thayeri", 🐭 🦎 🥚 85cm

Lampropeltis mexicana "greeri", 🐁 🦎 🌀 85cm

Lampropeltis mexicana "greeri", 🐀 🦎 🥚 85cm

Lampropeltis triangulum triangulum, 70cm

Lampropeltis triangulum annulata, 🐭 🦎 🥚 70cm
Lampropeltis triangulum campbelli, 🐭 🦎 🥚 85cm

Lampropeltis triangulum annulata, 70cm

155

Lampropeltis triangulum "temporalis", 🐁 🦎 ♂ 70cm
Lampropeltis triangulum "temporalis", 🐁 🦎 ♂ 70cm

Lampropeltis triangulum elapsoides, 🐍 🦎 🐚 50cm
Lampropeltis triangulum elapsoides, 🐍 🦎 🐚 50cm

Lampropeltis triangulum sinaloae, 120cm

Lampropeltis triangulum sinaloae, 🐁 🌙 120cm

Lampropeltis triangulum syspila, 🐸 🦎 ☾ 100cm
Lampropeltis triangulum syspila, 🐸 🦎 ☾ 100cm

Lampropeltis triangulum campbelli, 🐀 🦎 ✋ 85cm
Lampropeltis triangulum conanti, 🐀 🦎 ✋ 110cm

Lampropeltis triangulum hondurensis, 🐍 🦎 ♂ 110cm
Lampropeltis pyromelana knoblochi, 🐍 🦎 ♂ 90cm

Lampropeltis getula californiae, 🐭 🦎 ❄ 150cm
Lampropeltis triangulum arcifera, 🐭 🦎 ❄ 100cm

163

Lampropeltis zonata, 70cm

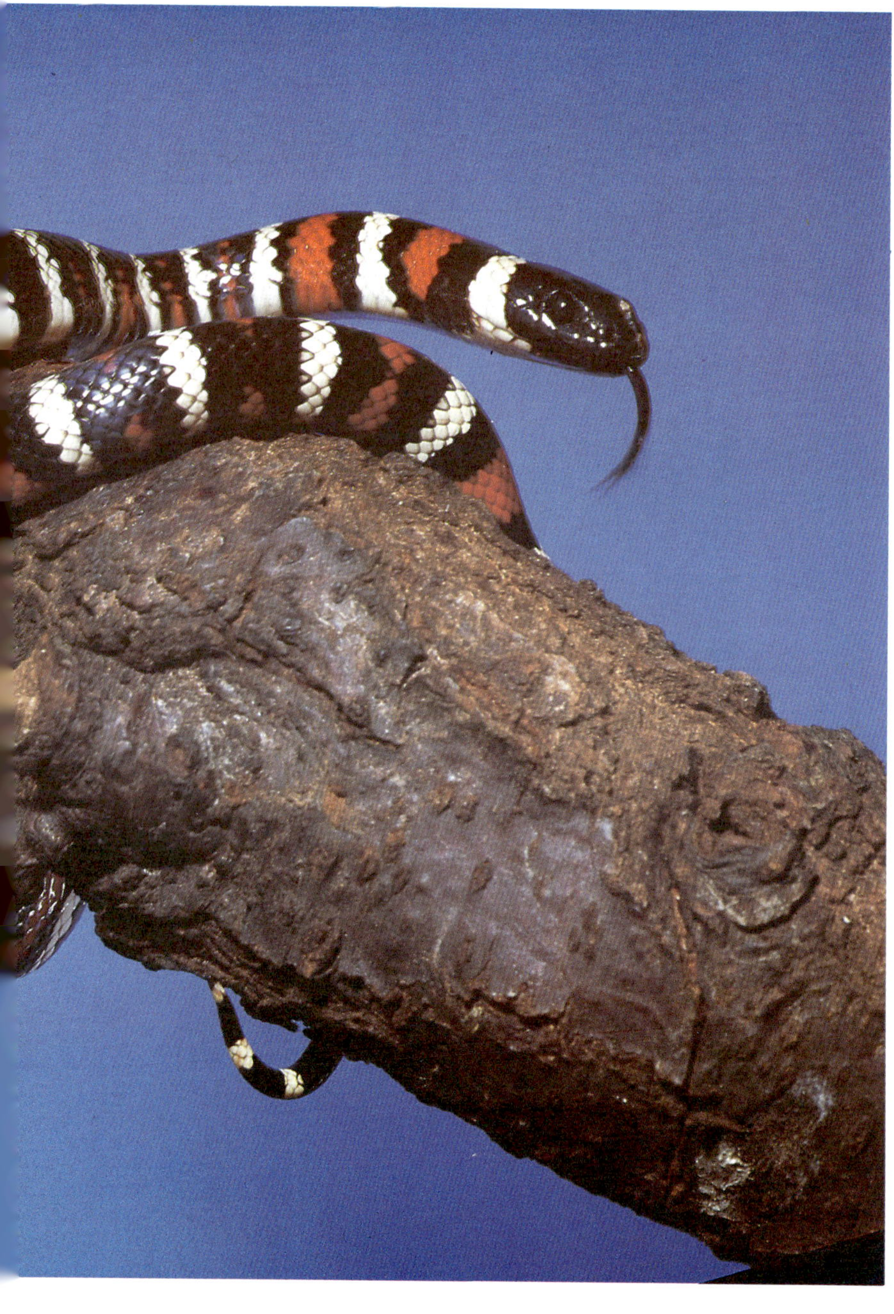

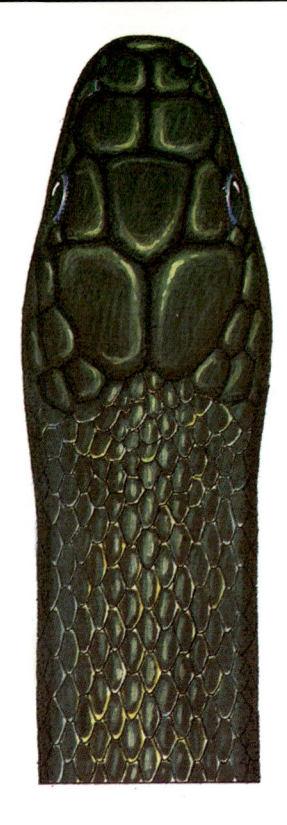

Lampropeltis getula nigrita, 75cm

Lampropeltis calligaster calligaster, 95cm

Lampropeltis calligaster rhombomaculata, 🐸🦎🥚 90cm

Lampropeltis calligaster occipitolineata, 🐸🦎🥚 90cm

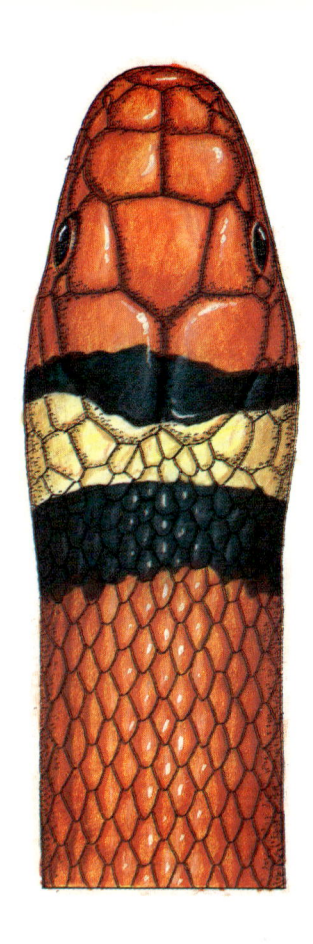

Lampropeltis triangulum elapsoides,
🐁 🦎 🥚 50cm

Lampropeltis alterna,
🐁 🦎 🥚 120cm

168

Lampropeltis getula splendida, 110cm

Lampropeltis mexicana, 85cm

Lampropeltis getula californiae, 150cm

Lampropeltis getula californiae, 150cm

Lampropeltis getula floridana, 150cm

Lampropeltis getula floridana, 150cm

Lampropeltis getula getula, 🐀 🦎 🥚 140cm

Lampropeltis triangulum andesiana, 🐀 🦎 🥚 130cm

Lampropeltis getula holbrooki, 115cm

Lampropeltis triangulum nelsoni, 100cm

Lampropeltis getula niger,
🐁 🐍 110cm

Lampropeltis mexicana "greeri",
🐁 🦎 🐍 85cm

Lampropeltis pyromelana infralabialis, 🐁 🦎 🥚 85cm

Lampropeltis pyromelana knoblochi, 🐁 🦎 🥚 90cm

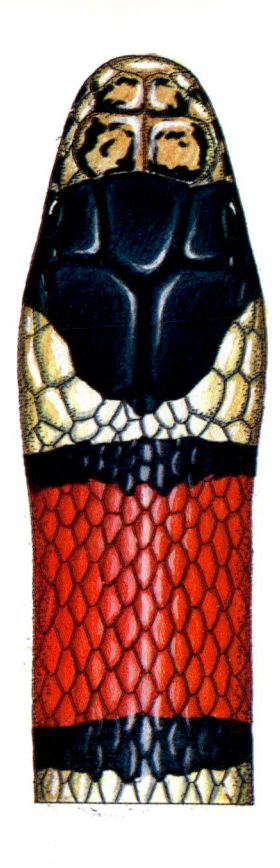

Lampropeltis pyromelana pyromelana, 🐢🦎🌀 95cm

Lampropeltis pyromelana woodini, 🐢🦎🌀 100cm

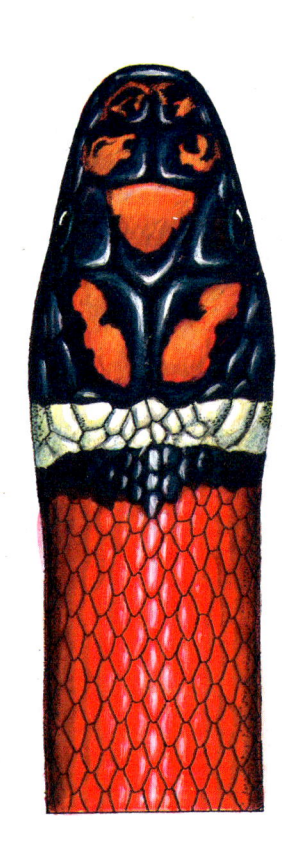

Lampropeltis ruthveni, 75cm

Lampropeltis triangulum abnorma, 140cm

Lampropeltis triangulum amaura, 50cm

Lampropeltis triangulum annulata, 70cm

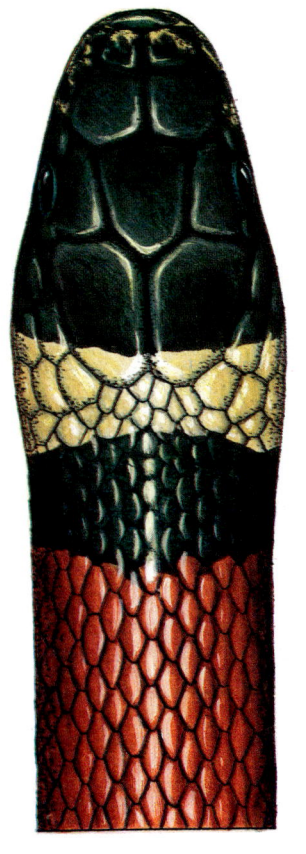

178

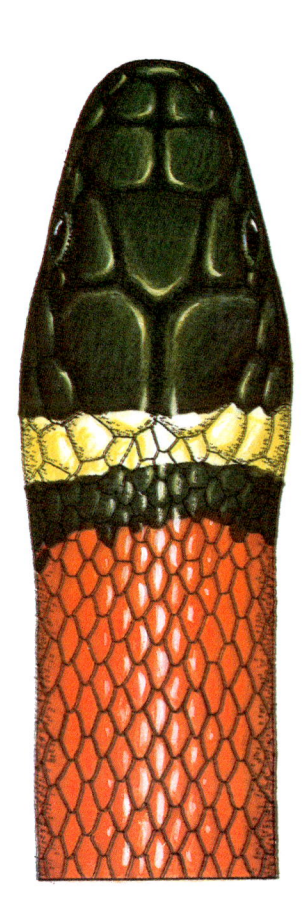

Lampropeltis triangulum arcifera,
🐸 🦎 🥚 100cm

Lampropeltis triangulum blanchardi,
🐸 🦎 🥚 100cm

Lampropeltis triangulum campbelli, 🐭 🦎 🥚 85cm

Lampropeltis triangulum celaenops, 🐭 🦎 🥚 55cm

Lampropeltis triangulum conanti, 110cm

Lampropeltis triangulum dixoni, 100cm

Lampropeltis triangulum gaigeae,
🐀 🦎 ⟲ 145cm

Lampropeltis triangulum gentilis,
🐀 🦎 ⟲ 85cm

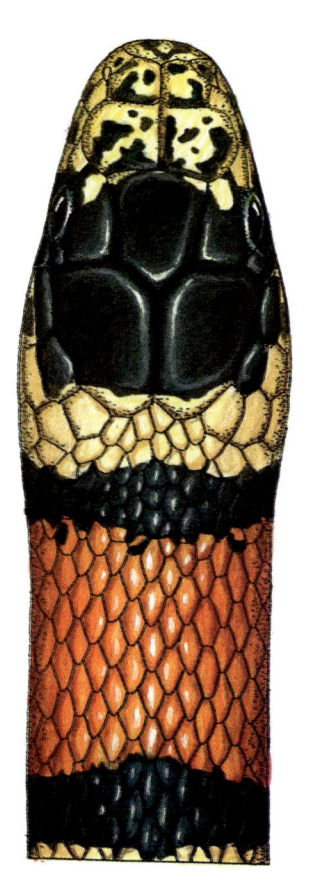

Lampropeltis triangulum hondurensis, 115cm

Lampropeltis triangulum micropholis, 175cm

Lampropeltis triangulum multistrata, 🐸🦎☀ 70cm

Lampropeltis triangulum oligozona, 🐸🦎☀ 100cm

Lampropeltis triangulum polyzona, 145cm

Lampropeltis triangulum sinaloae, 115cm

Lampropeltis triangulum smithi,
100cm

Lampropeltis triangulum stuarti,
110cm

Lampropeltis triangulum syspila, 65cm

Lampropeltis triangulum taylori, 70cm

Lampropeltis triangulum triangulum, 110cm

Lampropeltis zonata agalma 70cm

Lampropeltis zonata herrerae, 🐸 🦎 ☌ 70cm

Lampropeltis zonata multicincta, 🐸 🦎 ☌ 85cm

Lampropeltis zonata multifasciata, 🐸🦎 ♂ 85cm

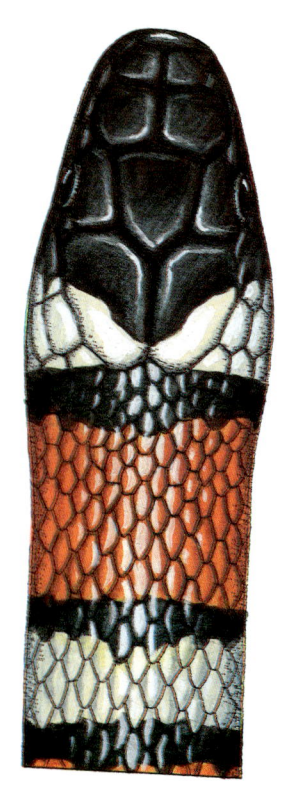

Lampropeltis zonata parvirubra, 🐸🦎 ♂ 95cm

Lampropeltis zonata pulchra, 85cm

Lampropeltis zonata zonata, 85cm

Lampropeltis zonata multifasciata, 🐸 🦎 ☉ 85cm
Lampropeltis zonata, 🐸 🦎 ☉ 85cm

Leptodrymus pulcherrimus, 🐭 🦎 ⚥ 120cm
Leptophis mexicanus, 🐦 🐸 ⚥ 130cm

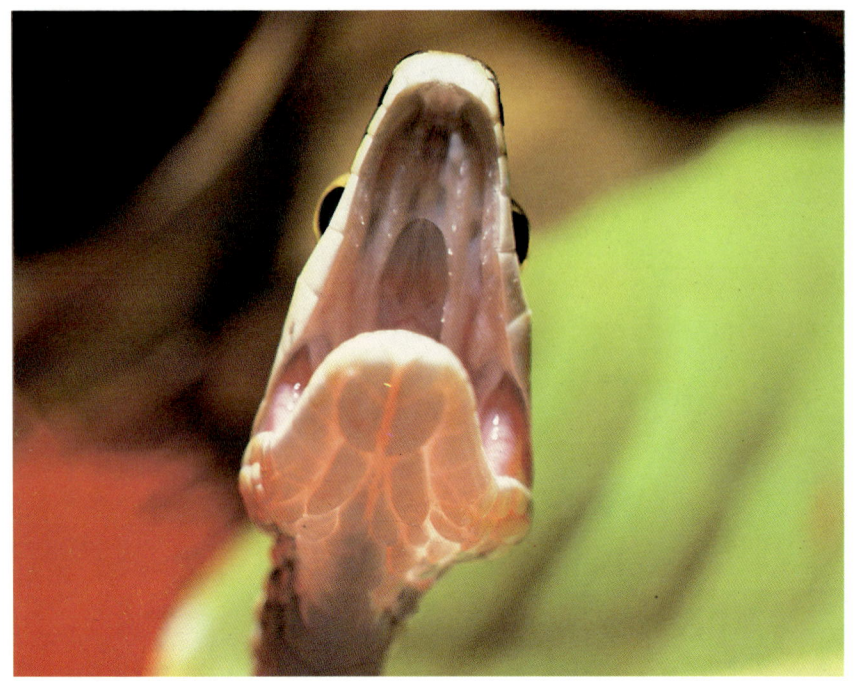

Leptophis mexicanus, 🐦 🐸 🥚 130cm
Leptophis diplotrophis, 🐦 🐸 🥚 130cm

Leptophis ahaetulla, 🐦 🐸 🌿 130cm

Leptophis ahaetulla, 🐦 🐸 🥚 130cm
Leptophis ahaetulla, 🐦 🐸 🥚 130cm

Lytorhynchus maynardi, 🦎 ● ☽ 40cm
Lytorhynchus ridgewayi, 🦎 ● ☽ 40cm

Masticophis bilineatus, 160cm

Masticophis flagellum, 225cm

Masticophis flagellum flagellum, 🐍🦎🥚 225cm
Masticophis flagellum testaceus, 🐍🦎🥚 225cm

Masticophis lateralis euryxanthus, 140cm
Masticophis flagellum piceus, 225cm

Masticophis flagellum testaceus, 225cm
Masticophis taeniatus schotti, 170cm

Masticophis lateralis lateralis, 🐀 🦎 ☽ 140cm
Masticophis flagellum piceus, 🐀 🦎 ☽ 225cm

Masticophis bilineatus, 160cm
Mastigodryas bifossatus, 140cm

Opheodrys vernalis, ✽ ♂ 45cm

Opheodrys vernalis, ✽ ♂ 45cm

Opheodrys aestivus, ✳ ∫ 70cm

Opheodrys vernalis, 🕷 🐍 45cm
Philothamnus hoplogaster, 🐸 🦎 🐍 100cm

Philothamnus semivariegatus, 🐦 🦎 🥚 100cm
Phyllorhynchus decurtatus perkinsi, 🦎 ● 🥚 40cm

Phyllorhynchus browni lucidus, 🦎 ● ☾ 40cm
Phyllorhynchus browni, 🦎 ● ☾ 40cm

Phyllorhynchus decurtatus, 🦎 ● 🥚 40cm
Pituophis melanoleucus melanoleucus, 🐢 🐦 🥚 160cm

Pituophis catenifer catenifer, 🐢 🐦 ☾ 275cm

Pituophis melanoleucus mugitus, 140cm
Pituophis melanoleucus melanoleucus, 160cm

Pituophis melanoleucus melanoleucus, 🐢 🐦 ☾ 160cm
Pituophis ruthveni, 🐢 🐦 ☾ 140cm

213

Pituophis melanoleucus melanoleucus, 🐢 🐦 🥚 160cm
Pituophis melanoleucus melanoleucus, 🐢 🐦 🥚 160cm

Pituophis catenifer affinis, 🐢 🐦 ⚪ 260cm

Pituophis sayi, 🐢 🐦 ⚪ 180cm

Pituophis deppei deppei, 🐀 🐦 💧 150cm
Pituophis deppei jani, 🐀 🐦 💧 150cm

Pituophis melanoleucus lodingi, 160cm

Pituophis melanoleucus lodingi, 160cm

Pituophis melanoleucus mugitus, 140cm
Pituophis sayi, 180cm

Prosymna frontalis, ● 🥚 35cm

Prosymna bivittata, ● 🥚 35cm

Pseudaspis cana, 120cm

Pseudaspis cana, 120cm

Pseustes poecilonotus, 🐟 ✋ 270cm
Pseustes sulphureus, 🐟 ✋ 270cm

Ptyas korros, 🐸 🦎 ☌ 350cm

Ptyas mucosus, 🐸 🦎 ☌ 350cm

Ptyas korros, 🐀 🦎 ☽ 350cm

Ptyas korros, 🐀 🦎 ☽ 350cm

Rhinocheilus lecontei tessellatus, 🦎 ● ♂ 75cm
Rhinocheilus lecontei tessellatus, 🦎 ● ♂ 75cm

224

Rhinocheilus lecontei antonii, 🐢 ♂ 75cm
Rhinocheilus lecontei tessellatus, 🦎 ● ♂ 75cm

Salvadora mexicanum, 🦎 ● 🥚 85cm
Salvadora grahamiae grahamiae, 🦎 ● 🥚 115cm

226

Salvadora deserticola, 🦎 ● ◡ 75cm
Salvadora grahamiae grahamiae, 🦎 ● ◡ 115cm

Salvadora deserticola, 🦎 ● 🐍 75cm

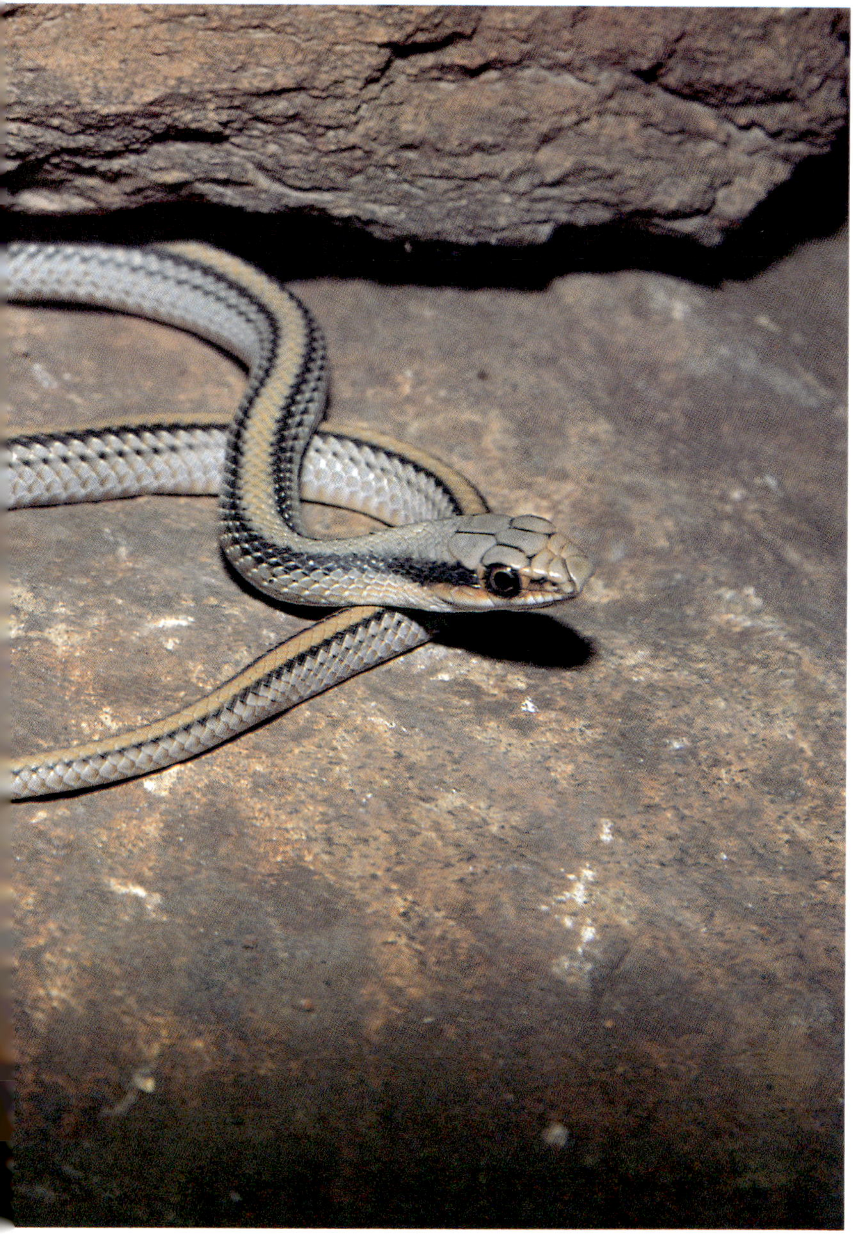

Spalerosophis diadema, 140cm
Spalerosophis arenarius, 140cm

230

Spilotes pullatus, 🐁 🐦 🥚 240cm

Spilotes pullatus, 🐁 🐦 🥚 240cm

Stilosoma extenuatum, 45cm

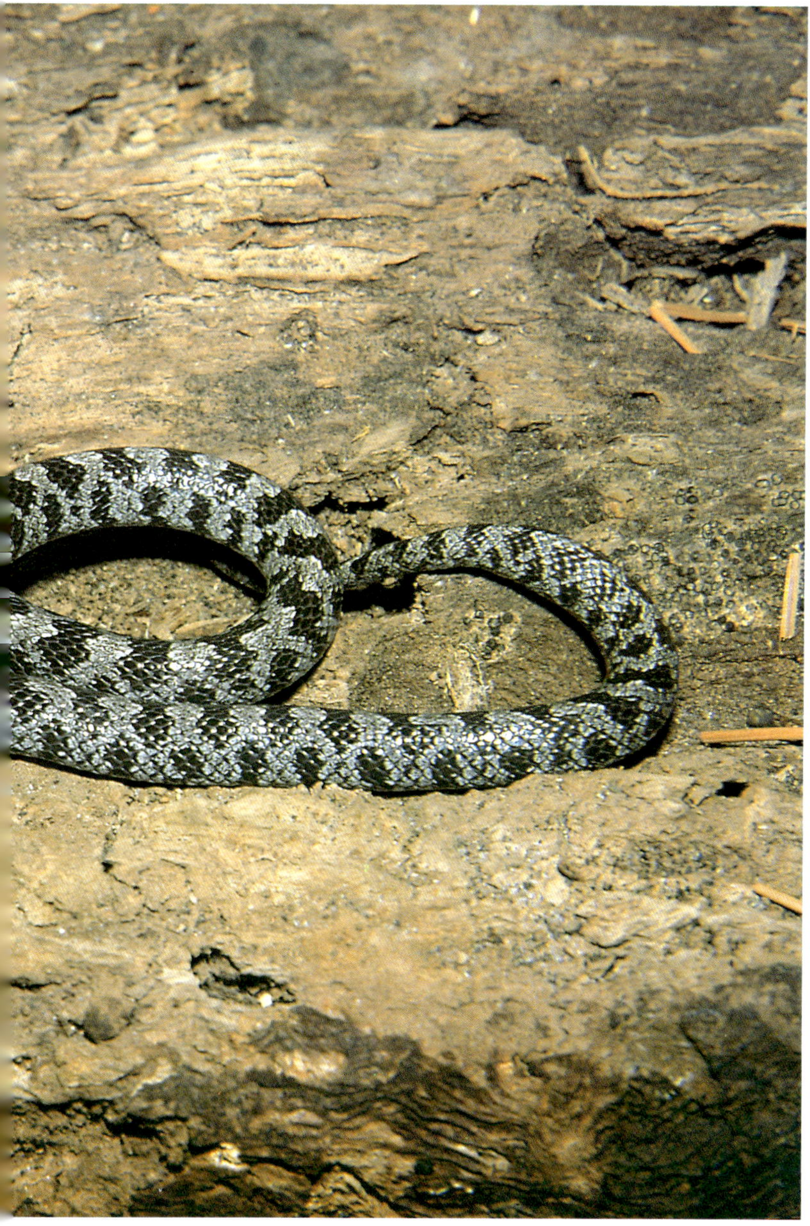

Zaocys dhumnades, 🦎 ♂ 200cm   *Zaocys dhumnades,* 🦎 ♀ 200cm

Dasypeltis scabra, ● ♂ 80cm

Dasypeltis scabra, ● ♂ 80cm

Dasypeltis inornata, ● ⚲ 80cm

Boaedon fuliginosus, 🦎 🐸 ⚲ 75cm

Boaedon fuliginosus, 🦎 🐸 🥚 75cm
Dinodon rufozonatum, 🐟 🐸 🥚 100cm

Farancia abacura abacura, 🐸 ♂ 130cm
Farancia abacura abacura, 🐸 ♂ 130cm

Farancia abacura abacura, 🐟 ☾ 130cm
　　　　　Farancia erytrogramma erytrogramma, 🐟 🐟 ☾ 115cm

Farancia erytrogramma erytrogramma, 115cm

Farancia erytrogramma erytrogramma, 🐟 🐸 🥚 115cm

Lamprophis aurora, 70cm

Lamprophis aurora, 70cm

Lamprophis guttatus, 🦎 ♂ 60cm

Lamprophis fiski, 🦎 ♂ 35cm

243

Liopholidophis sexlineatus, 🐦 🥚 80cm

Liopholidophis stumpffi, 🐦 🥚 80cm

Lycodon striatus bicolor, 70cm

Lycodon aulicus, 🦎 ✳ 🥚 70cm

Lycodon subcinctus, 🦎 ✳ 🥚 70cm

Lycodonomorphus whytii, 🐟 🐸 🥚 100cm

Lycophidion capense, 🦎 🥚 45cm

Mehelya capensis, 🐟 🦎 🐍 130cm

Mehelya crossi, 🐟 🦎 🐍 95cm

Ninia sebae, 🦎 🌙 50cm

Oligodon taeniatus, 🦎 ● 🌙 60cm

249

Oligodon taeniolatus, 🦎 ● ✋ 60cm

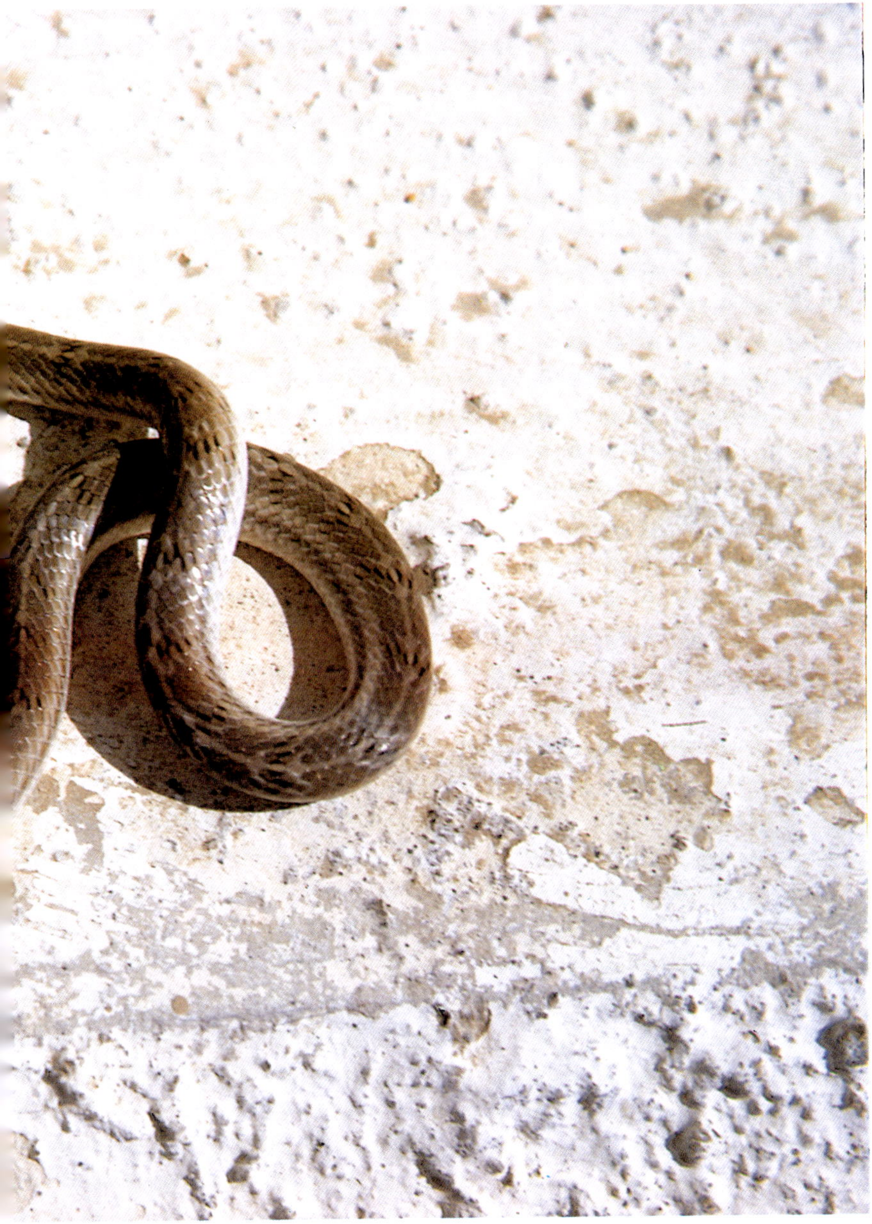

Rhynchocalamus melanocephalus, 40cm

Amphiesma sp., 🐟 🐸 ☾ 75cm

Amphiesma stolata, 🐟 🐸 ☾ 75cm

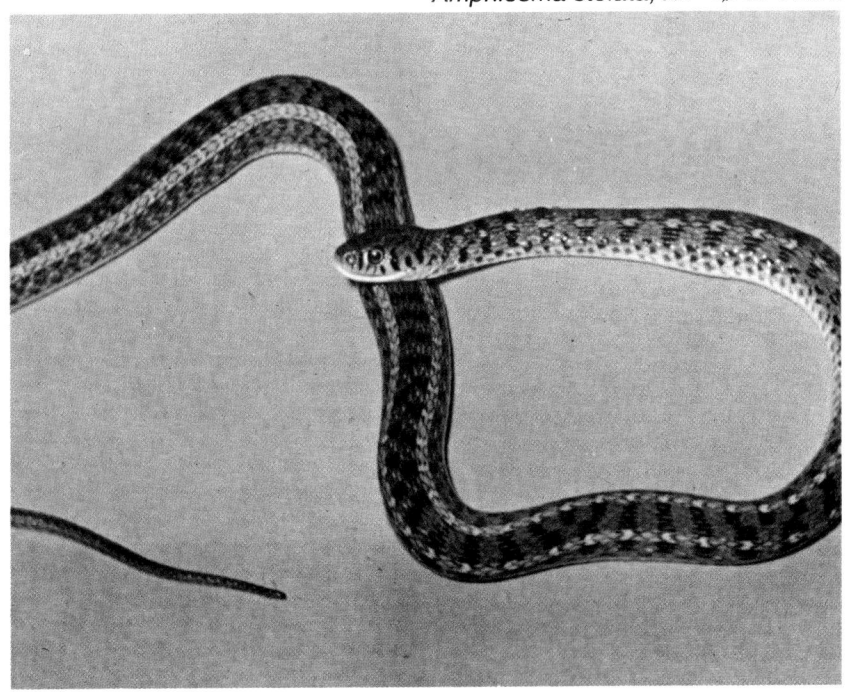

Carphophis amoenus amoenus, 🕷 🐍 30cm

Carphophis amoenus amoenus, 🕷 🐍 30cm

254

Diadophis punctatus, 🐾 ✳ 🐚 30cm

Diadophis punctatus regalis, 🦎 🕷 ☾ 45cm
Diadophis punctatus, 🦎 🕷 ☾ 30cm

256

Diadophis punctatus punctatus, 🦎 🕷 🌙 35cm
Diadophis punctatus pulchellus, 🦎 🕷 🌙 60cm

Diagophis punctatus arnyi, 🦎 🕷 🥚 60cm

Diadophis punctatus amabilis, 🦎 🕷 ☾ 65cm

Gonyosoma oxycephalum, 160cm

Grayia smithii, 200cm

Helicops leopardinus, 🐟 🐸 ☾ 100cm
Helicops carinicauda, 🐟 🐸 ☾ 100cm

Helicops sp., 🐟 🐸 ⌂ 100cm

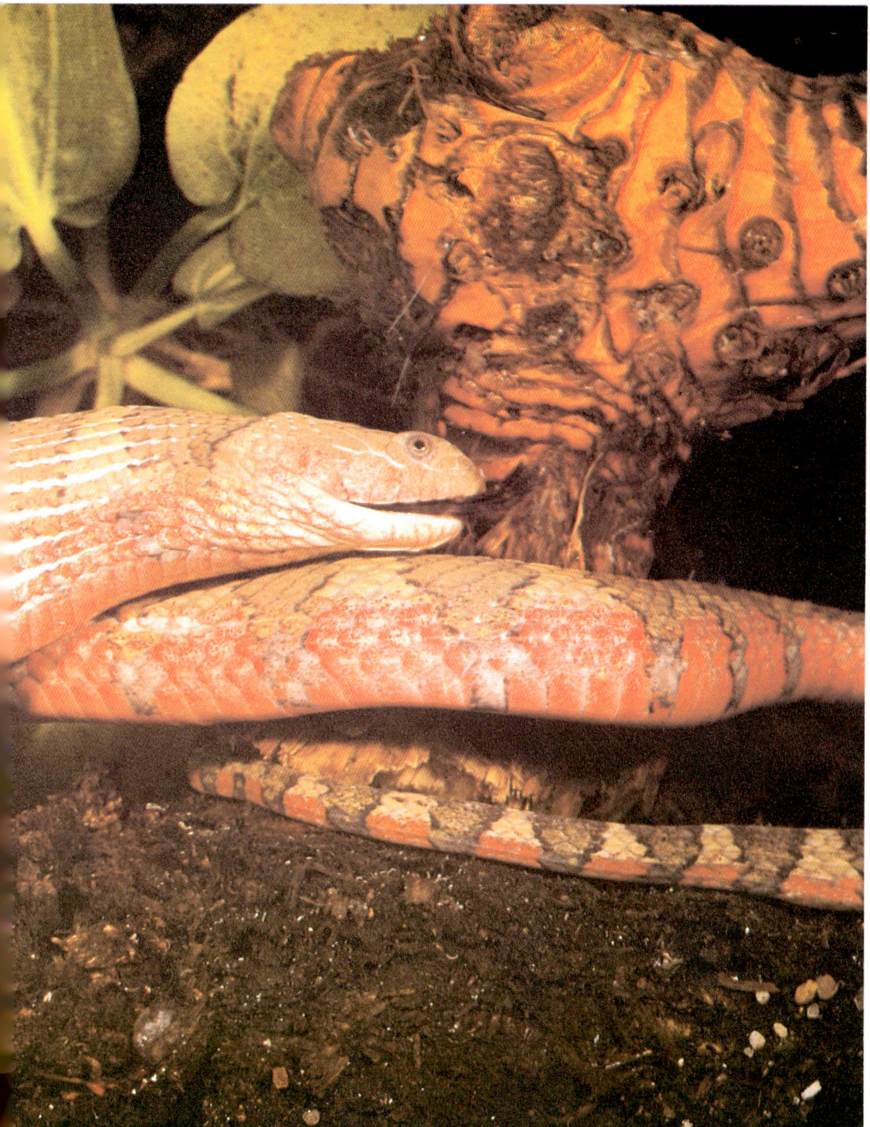

Natriciteres olivacea, 🐸 🐚 50cm

Natrix tessellata, 🐟 🐸 🐚 145cm

264

Natrix maura, 🐟 🐸 ⚬ 100cm

Natrix maura, 100cm

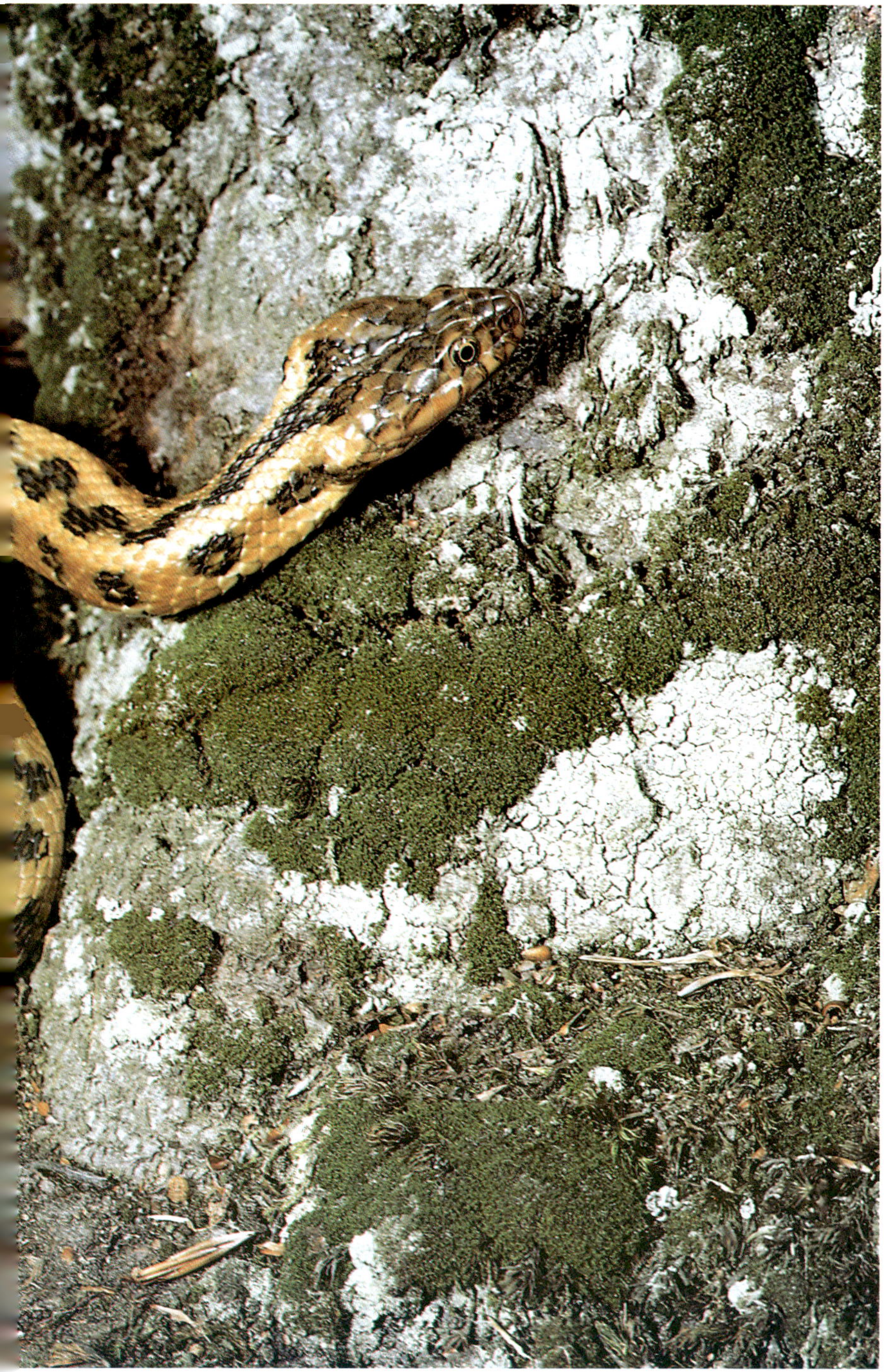

Natrix natrix helvetica, 🐟 🐟 💧 140cm

Natrix tessellata, 🐟 🐟 💧 145cm

Nerodia erythrogaster flavigaster, 🐟 🐸 〽 120cm
Nerodia fasciata confluens, 🐟 🐸 〽 105cm

Nerodia clarkii, 🐟 🐸 🌙 75cm

Nerodia cyclopion, 🐟 🐸 🌙 110cm

Nerodia fasciata confluens, 🐟 🐸 ☾ 105cm
Nerodia fasciata fasciata, 🐟 🐸 ☾ 105cm

271

Nerodia erythrogaster flavigaster, 🐟 🐸 〰 120cm

Nerodia fasciata pictiventris, 🐟 🐸 〰 105cm

Nerodia clarkii, 🐟 🐸 🐍 75cm
Nerodia erythrogaster erythrogaster, 🐟 🐸 🐍 120cm

Nerodia fasciata pictiventris, 🐟 🐸 🍌 105cm
Nerodia rhombifer rhombifer, 🐟 🐸 🍌 120cm

Nerodia fasciata fasciata, 105cm

Pliocercus sp., 65cm

Regina rigida rigida, 🐟 🐛 🌙 60cm

Regina septemvittata, 🐟 🌙 60cm

Regina grahami, 🐟 🐸 🐛 70cm

Rhabdophis subminiatus, 🐸 ☾ ☠ 140cm
Rhabdophis tigrina, 🐸 ☾ ☠ 140cm

Seminatrix pygaea cyclas, 🐸 🕷 ⌇ 30cm
Seminatrix pygaea, 🐸 🕷 ⌇ 30cm

Storeria dekayi, 🕷 🐁 35cm

Storeria occipitomaculata, 🕷 🐁 30cm

Storeria occipitomaculata, ✳ 🍂 30cm

Storeria dekayi, ✳ 🍂 35cm

Storeria dekayi victa, 🕷 🐌 30cm

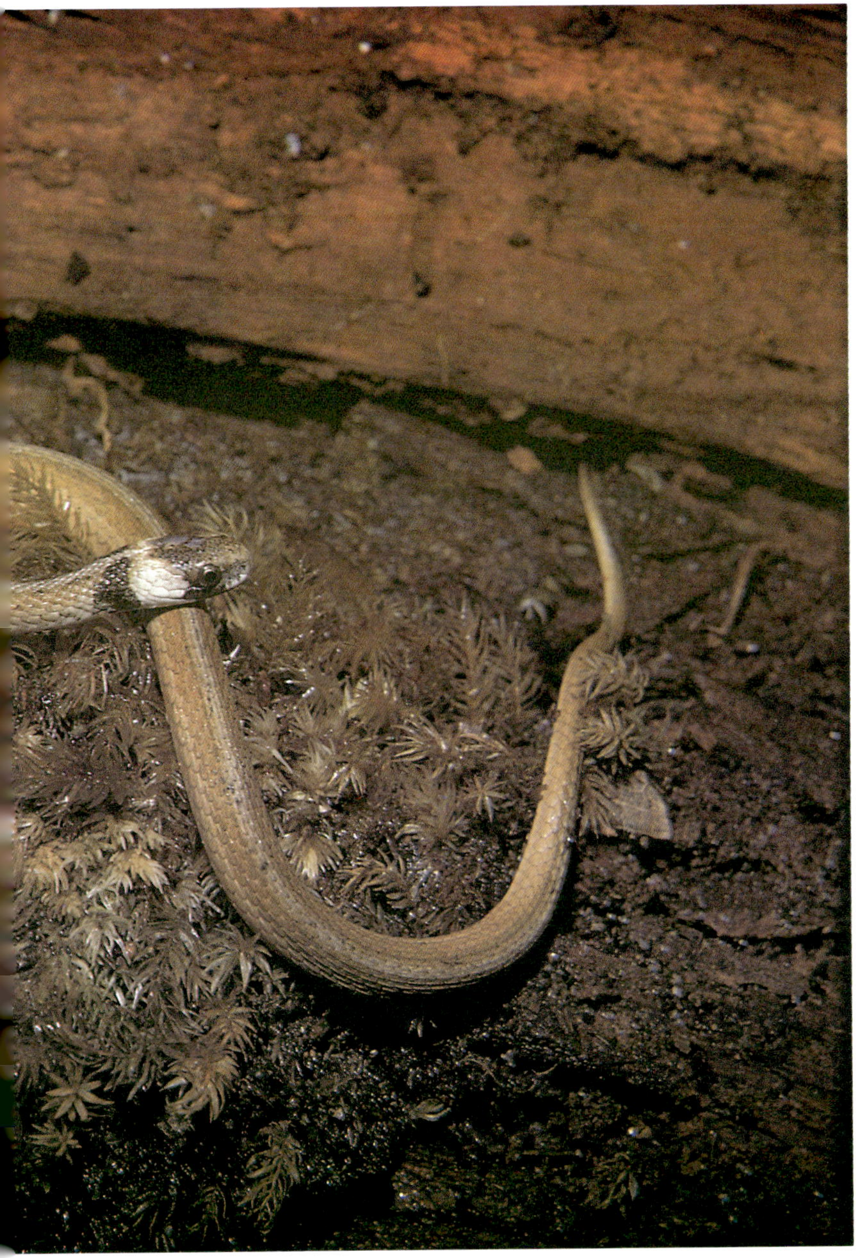

Thamnophis butleri, 🐟 🕷 🐌 50cm

Thamnophis sauritus nitae, 🐟 🕷 🐌 65cm

284

Thamnophis sirtalis tetrataenia, 🐟 🕷 🐸 65cm
Thamnophis sirtalis parietalis, 🐟 🕷 🐸 65cm

Thamnophis sirtalis, 🐟 🕷 👅 65cm

Thamnophis brachystoma, 🐟 🕷 🐛 45cm

Thamnophis sirtalis sirtalis, 🐟 🕷 🐁 65cm
Thamnophis radix, 🐟 🕷 🐁 70cm

288

Thamnophis sirtalis concinnus, 🐟 🕷 🐛 65cm
Thamnophis marcianus marcianus, 🐟 🕷 🐛 60cm

289

Thamnophis sirtalis sirtalis, 65cm

Thamnophis sirtalis similis, 🐟 🕷 🐸 65cm

Thamnophis proximus proximus, 🐟 🕷 🐛 75cm
Thamnophis proximus rubrilineatus, 🐟 🕷 🐛 75cm

Thamnophis sirtalis, 🐟 🕷 🪱 65cm

Thamnophis radix, 🐟 🕷 🪱 70cm

Thamnophis sirtalis similis, 65cm
Thamnophis hammondii, 65cm

Thamnophis marcianus, 🐟 🕷 🐸 60cm

Thamnophis marcianus, 🐟 🕷 🐸 60cm

Thamnophis sauritus sackeni, 🐟 🕷 🌀 65cm

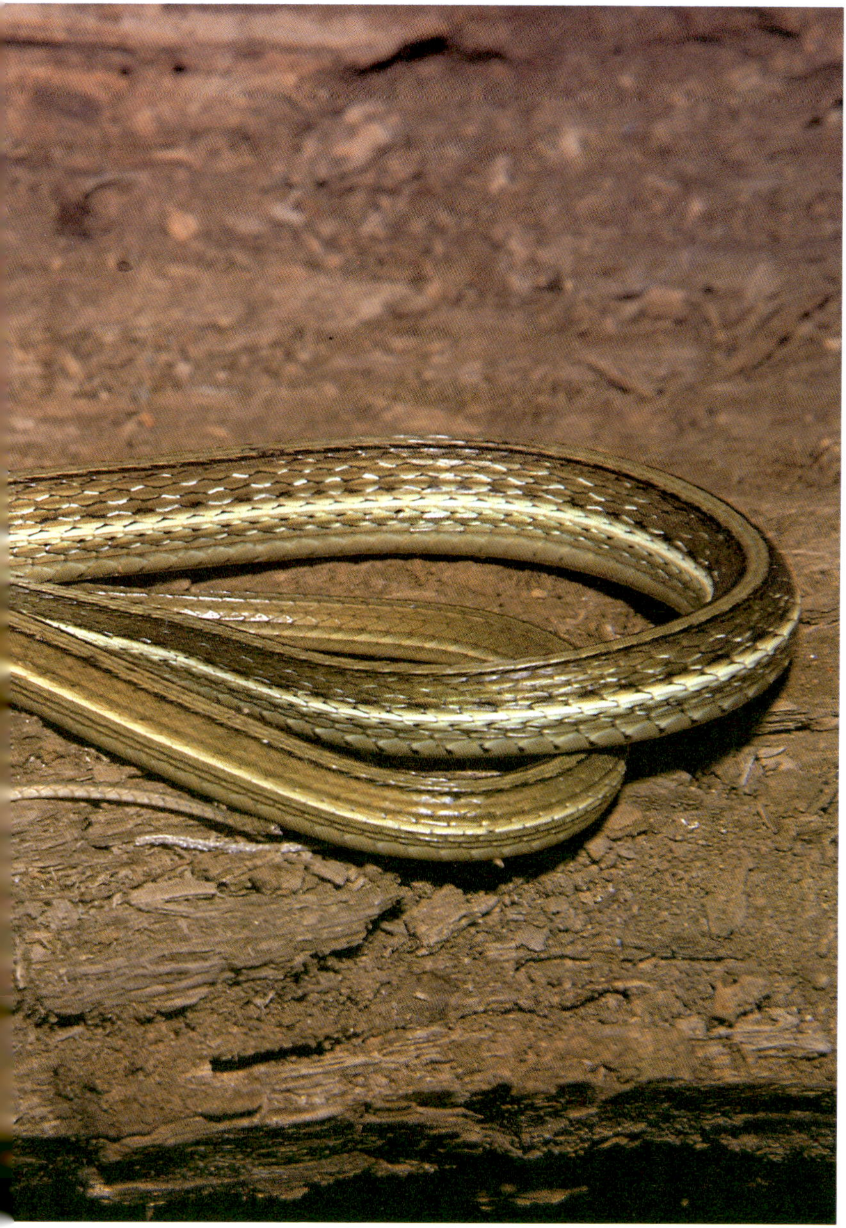

Thamnophis sirtalis sirtalis, 🐟 🕷 🐌 65cm
Thamnophis sirtalis sirtalis, 🐟 🕷 🐌 65cm

Thamnophis sirtalis tetrataenia, 🐟 🕷 🐛 65cm
Thamnophis sirtalis, 🐟 🕷 🐛 65cm

299

Thamnophis ordinoides, 🐟 🕷 🐌 95cm
Thamnophis eques megalops, 🐟 🕷 🐌 100cm

Thamnophis cyrtopsis, 🐟 🕷 🐸 105cm
Thamnophis sauritus nitae, 🐟 🕷 🐸 65cm

Tropiclonion lineatum texanum, ✳ ✍ 35cm

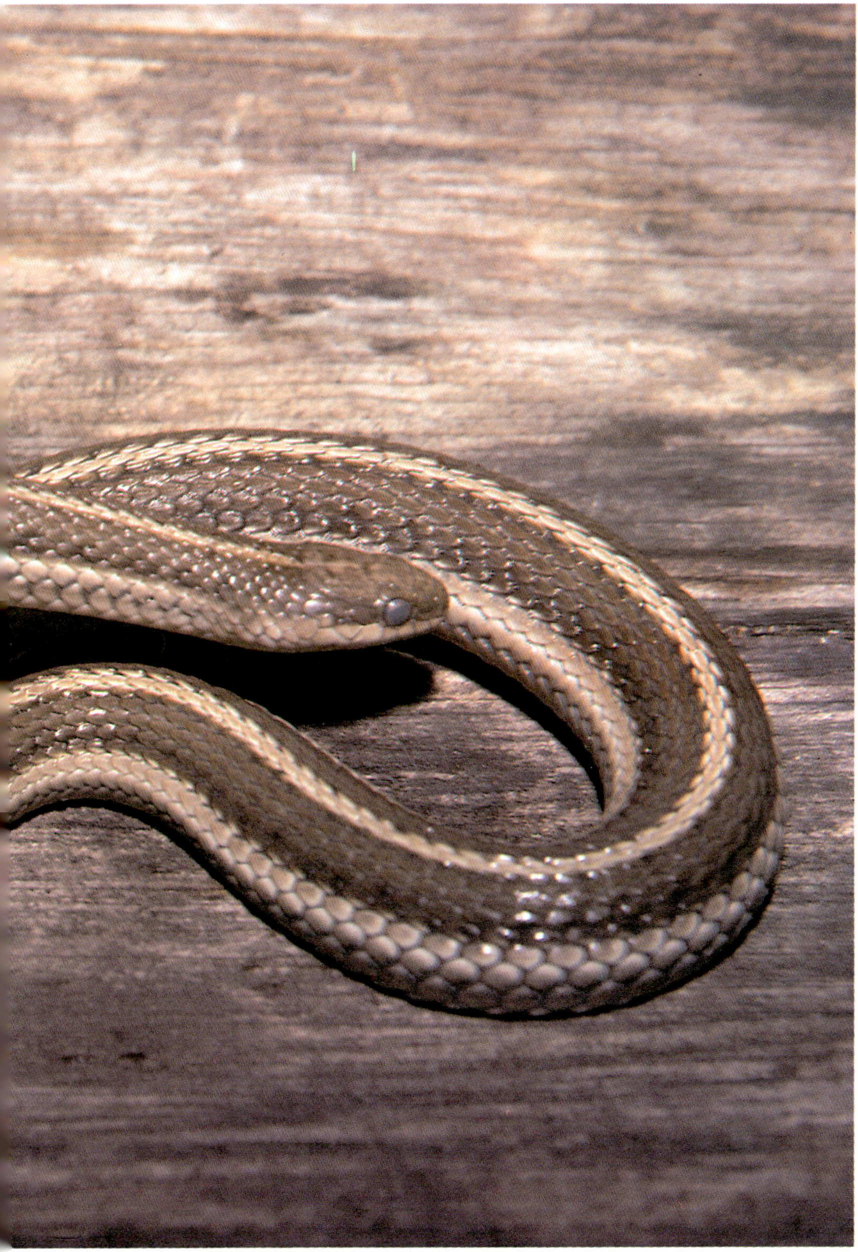

Tropiclonion lineatum lineatum, ✲ 🐌 35cm

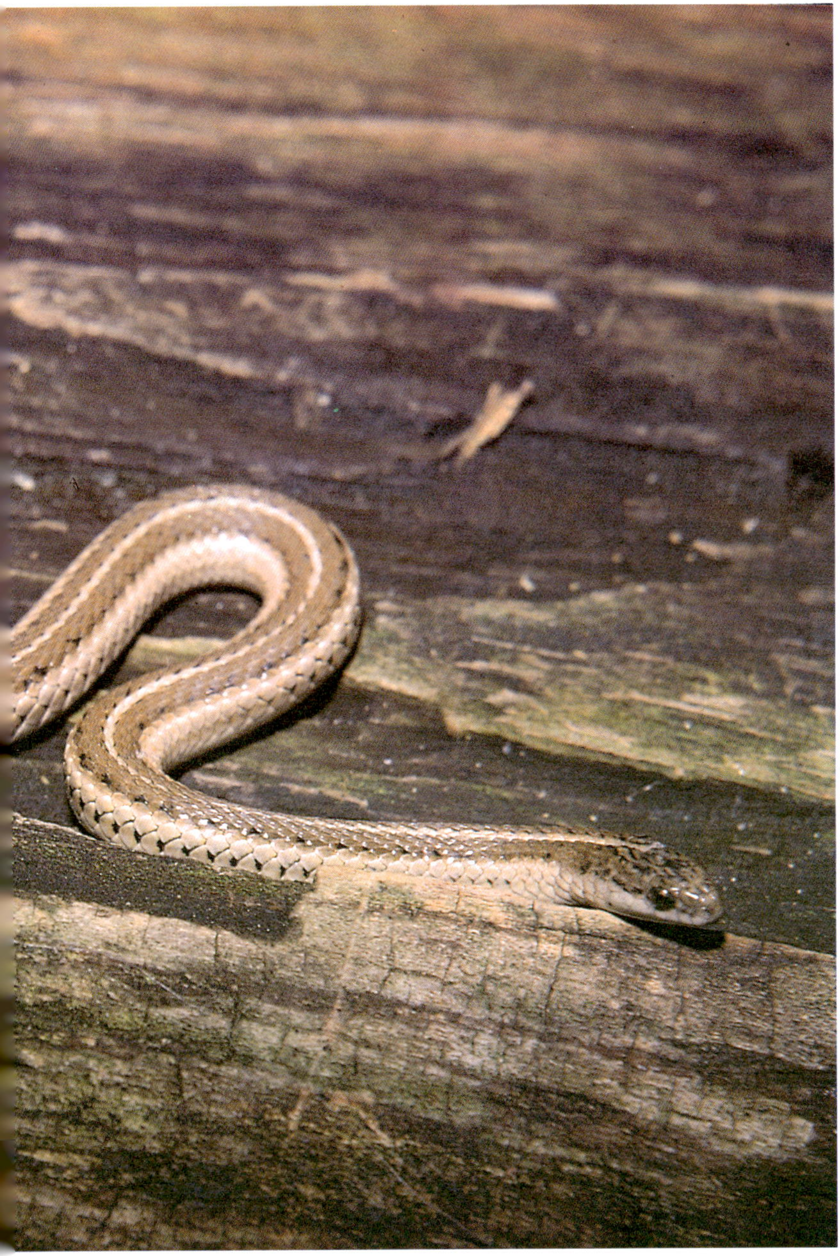

Virginia valeriae elegans, 25cm

Virginia valeriae, 25cm

Virginia valeriae, ✳🝰 25cm

Xenochrophis piscator, 🐟 🐚 125cm

Xenochrophis piscator, 🐟 🐚 125cm

Cerberus rhynchops, 🐟 🐛 85cm

Enhydris chinensis, 🐟 🐟 🐛 145cm

309

Enhydris plumbea, 🐟 🐸 🐍 145cm

Enhydris jagori, 🐟 🐸 🐍 145cm

Erpeton tentaculatus, 🐟 🥐 80cm

Erpeton tentaculatus, 🐟 🥐 80cm

Ahaetulla prasina, 180cm

Ahaetulla nasuta, 160cm

Amplorhinus multimaculata, 60cm
Boiga cyanea, 190cm

Boiga dendrophila, 250cm
Boiga irregularis, 200cm

314

Boiga ocellata, 🐢 🦎 🌙 ☠ 210cm
Boiga irregularis, 🐢 🦎 🌙 ☠ 200cm

Boiga dendrophila, 🐀 🦎 🥚 ☠ 250cm

Boiga atriceps, 🐁 🦎 💧 ☠ 195cm
Boiga cynodon, 🐁 🦎 💧 ☠ 170cm

Boiga fusca, 🐍🦎☾☠ 200cm
Boiga multomaculata, 🐍🦎☾☠ 210cm

Chrysopelea ornata, 🦎 🐸 ⟲ 130cm
Chrysopelea ornata, 🦎 🐸 ⟲ 130cm

319

Chrysopelea ornata, 130cm

Chrysopelea ornata, 130cm

Clelia rustica, 🐍 ♂ ☠ 250cm
Clelia occipitolutea, 🐍 ♂ ☠ 250cm

Clelia clelia, 🦎 🥚 ☠ 250cm

Clelia clelia clelia, 🦎 🥚 ☠ 250cm

Coniophanes imperialis, 🐁 🐸 🦆 ☠ 65cm

325

Coniophanes piceivittis, 🐟 🐸 ✋ ☠ 65cm
Crotaphopeltis hotamboeia, 🦎 🐸 ✋ 70cm

326

Dipsadoboa pulverulentus, 🐸 💤 ☠ 100cm

Dipsadoboa aulica, 🐸 💤 ☠ 100cm

Dispholidus typus, 🐀 🐦 ✋ ☠ 185cm
Dispholidus typus, 🐀 🐦 ✋ ☠ 185cm

Dispholidus typus, 🐭 🐦 🥚 ☠ 185cm

Dispholidus typus, 🐭 🐦 🥚 ☠ 185cm

Dispholidus typus, 🐁 🐦 ✋ ☠ 185cm
Elapomorphus bilineatus, 🦎 🦗 ✋ 25cm

330

Enulius flavitorques, ✳ 🜂 35cm
   Erythrolamprus mimus, 🦎 🜂 ☠ 100cm

Erythrolamprus bizonus, 🐾 ☁ ☠ 100cm

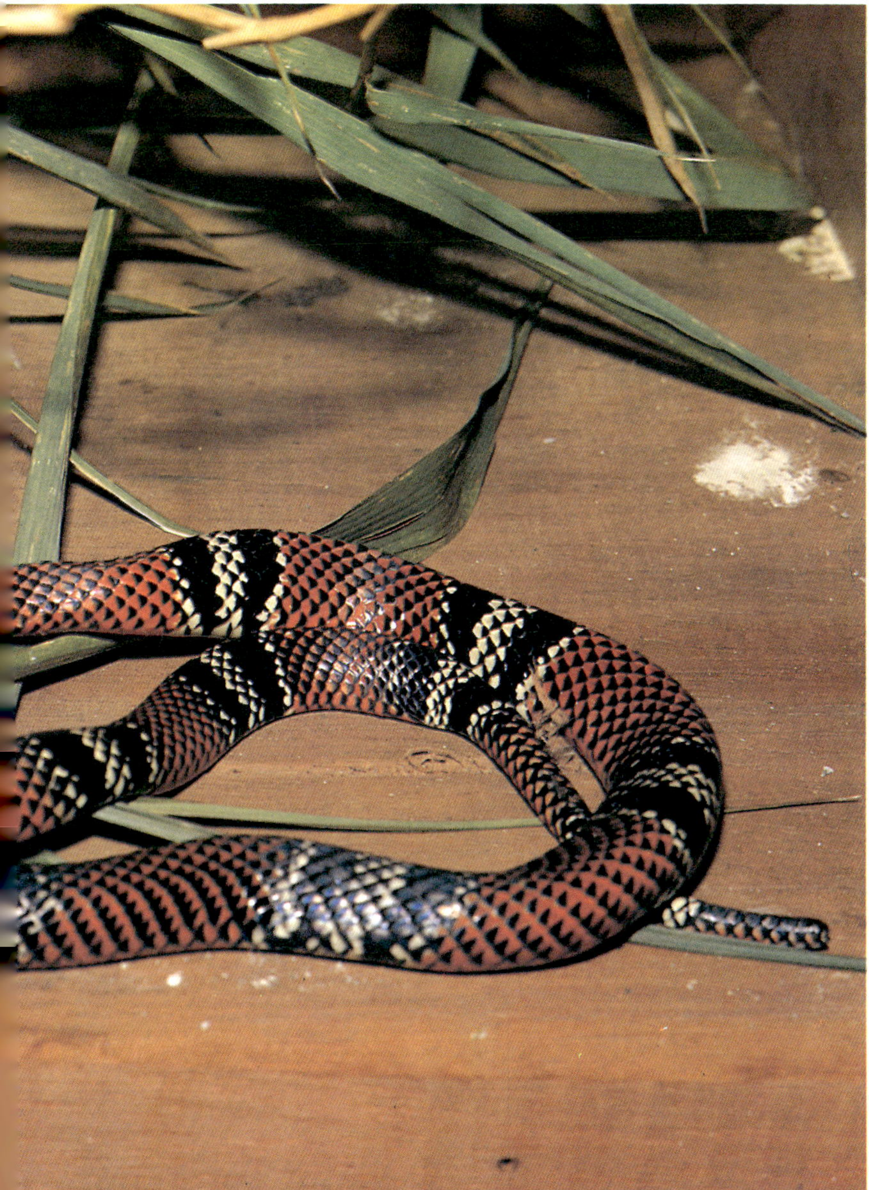

Erythrolamprus aesculapii, 🐾 ☾ ☠ 100cm

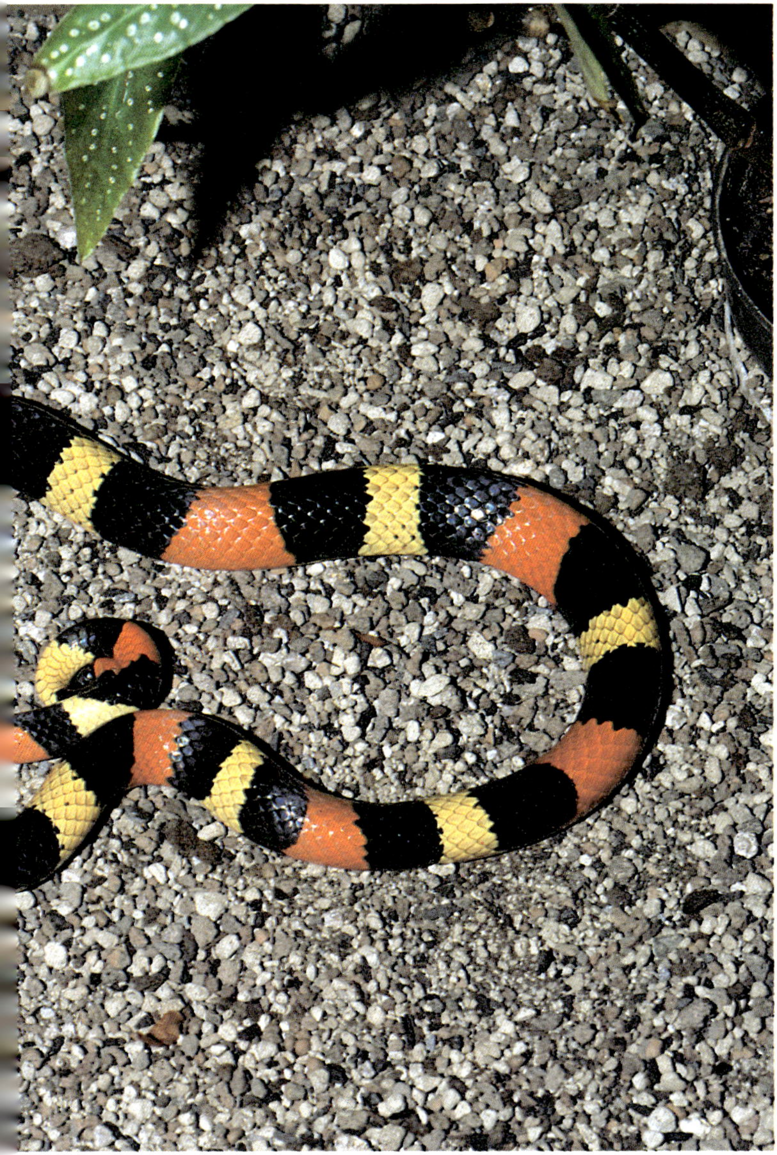

Erythrolamprus aesculapii, 🐕 ✋ ☠ 100cm
Erythrolamprus aesculapii, 🐕 ✋ ☠ 100cm

Ficimia streckeri, ✴ ♂ 40cm

Ficimia streckeri, ✴ ♂ 40cm

Gyalopion canum, 🦎 ✴ ☾ 35cm
Gyalopion quadrangularis, 🦎 ✴ ☾ 40cm

338

Hemirhagerrhis nototaenia, 🦎 🐸 ✋ ☠ 40cm
Hydrodynastes bicinctus, 🐸 ✋ 180cm

Hydrodynastes gigas, 🐸 🐁 180cm

Hydrodynastes gigas, 🐸 🐁 180cm

Hydrodynastes gigas, 🐸 ☾ 180cm

Hypsiglena torquata, 🦎 🐸 ☾ 65cm

Imantodes cenchoa, 🦎 🐸 ☾ ✻ 110cm

Imantodes cenchoa, 🦎 🐸 ☾ ✻ 110cm

Imantodes sp., 110cm

Leptodeira septentrionalis septentrionalis, 🐸 ☾ 60cm

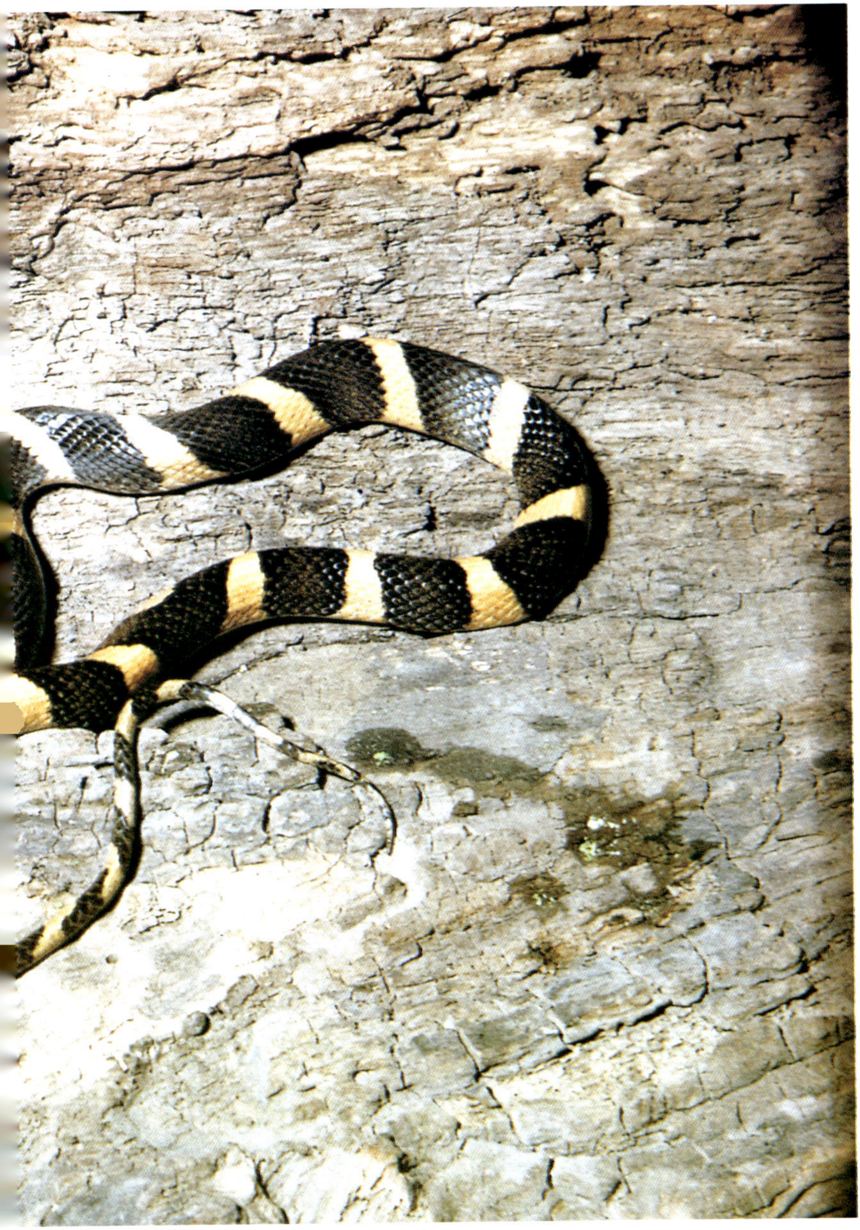

Leptodeira nigrofasciata, 🐸 ☾ 60cm

Leptodeira annulata, 🐸 ☾ 60cm

Leptodeira septentrionalis, 🐸 🜂 60cm
Leptodeira septentrionalis, 🐸 🜂 60cm

Leptodeira septentrionalis septentrionalis, 🐍 ♂ 60cm

Macroprotodon cucullatus, 🐍 ♂ 50cm

Madagascarophis colubrinus, 🐁 🦎 🐍 100cm

Madagascarophis colubrinus, 100cm
Malpolon moilensis, 170cm

350

Malpolon moilensis, 🐍 🦎 ☡ ☠ 170cm
Malpolon monspessulanus, 🐍 🦎 ☡ ☠ 170cm

Mimophis mahafalensis, 🐀 🦎 🥚 ☠ 100cm
Mimophis madagascariensis, 🐀 🦎 🥚 ☠ 100cm

352

Oxybelis aeneus, 🦎 🐸 ☝ ☠ 130cm
Oxybelis brevirostris, 🦎 🐸 ☝ ☠ 130cm

Oxybelis argenteus, 🦎 🐸 🥚 ☠ 130cm

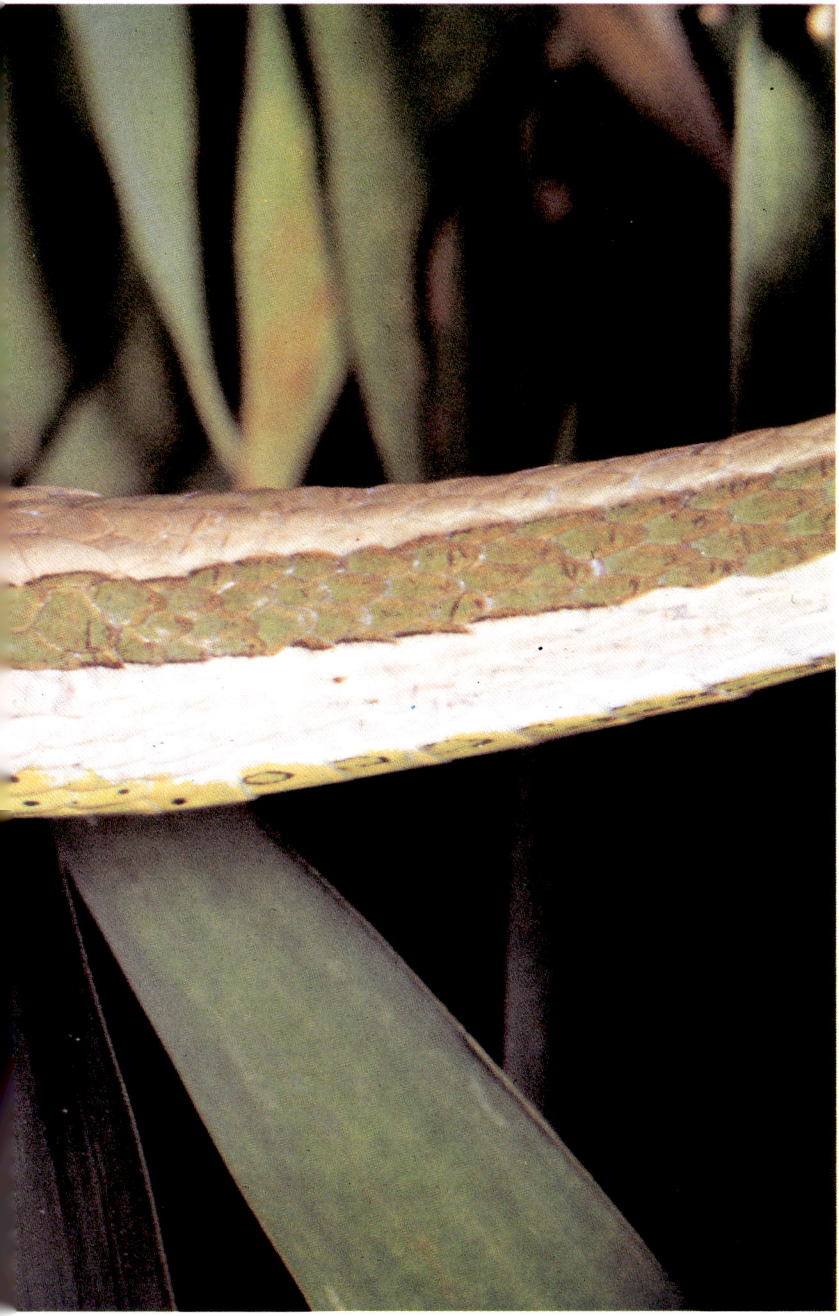

Oxybelis fulgidus, 🦎 🐸 🥚 ☠ 130cm

Oxybelis spp., 🦎 🐸 🥚 ☠ 130cm

Oxyrhopus rhombifer, 🦎 💤 ☠ 90cm

Oxyrhopus petolarius, 🦎 💤 ☠ 90cm

Oxyrhopus petolarius, 🐾 ☾ ☠ 90cm

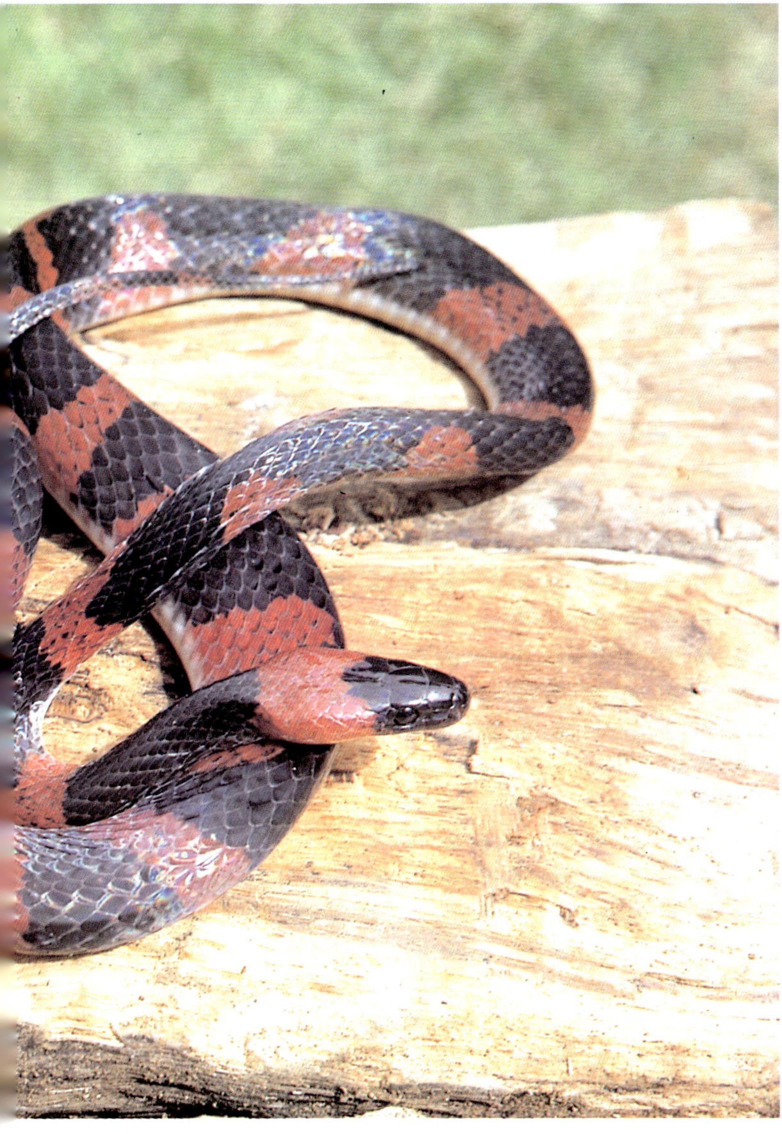

Oxyrhopus rhombifer, 🐾 ☾ ☠ 90cm
Oxyrhopus trigeminus, 🐾 ☾ ☠ 90cm

Phimophis guerini, 🦎 💤 ☠ 90cm

Phimophis guerini, 🦎 💤 ☠ 90cm

361

Psammodynastes pulverulentus, 🦎 🐸 🐍 ☠ 65cm

Psammophis jallae, 🐁 🦎 🥚 10 90cm

Psammophis sp., 🐍 🦎 🥚 ☠ 95cm
Psammophis leightoni namibensis, 🐍 🦎 🥚 ☠ 100cm

Psammophis leithi, 🐀 🦎 ☂ ☠ 95cm

Psammophis lineolatus, 🐢 🦎 ☁ ☠ 100cm
Psammophis lineolatus, 🐢 🦎 ☁ ☠ 100cm

Psammophis leightoni trinasalis, 🐍🦎☪☠ 100cm
Psammophis phillipsii, 🐍🦎☪☠ 130cm

Psammophylax rhombeatus, 🐍 🦎 ☁ ☠ 105cm
Psammophylax rhombeatus, 🐍 🦎 ☁ ☠ 105cm

Psammophylax tritaeniatus, 🐢🦎☾☠ 105cm
Psammophylax rhombeatus, 🐢🦎☾☠ 105cm

Psammophylax tritaeniatus, 🐁 🦎 ☠ ☠ 105cm
Pseudoboa neuwiedii, 🐁 🦎 ☠ 100cm

Pythonodipsas carinata, 🦎 🥚 60cm
Rhamphiophis oxyrhynchus, 🐸 🦎 🥚 140cm

Rhamphiophis oxyrhynchus, 🐸 🦎 ♂ 140cm

Rhamphiophis oxyrhynchus, 🐟 🦎 ✋ 140cm
Rhamphiophis oxyrhynchus, 🐟 🦎 ✋ 140cm

Rhinobothryum bovalli, 🐦 🦡 ♂ 160cm

Sonora aemula, ✴ 🜨 50cm

Sonora aemula, ✴ 🜨 50cm

Sonora semiannulata, ✴ 🐍 50cm

Sonora semiannulata, 🕷 🐌 50cm

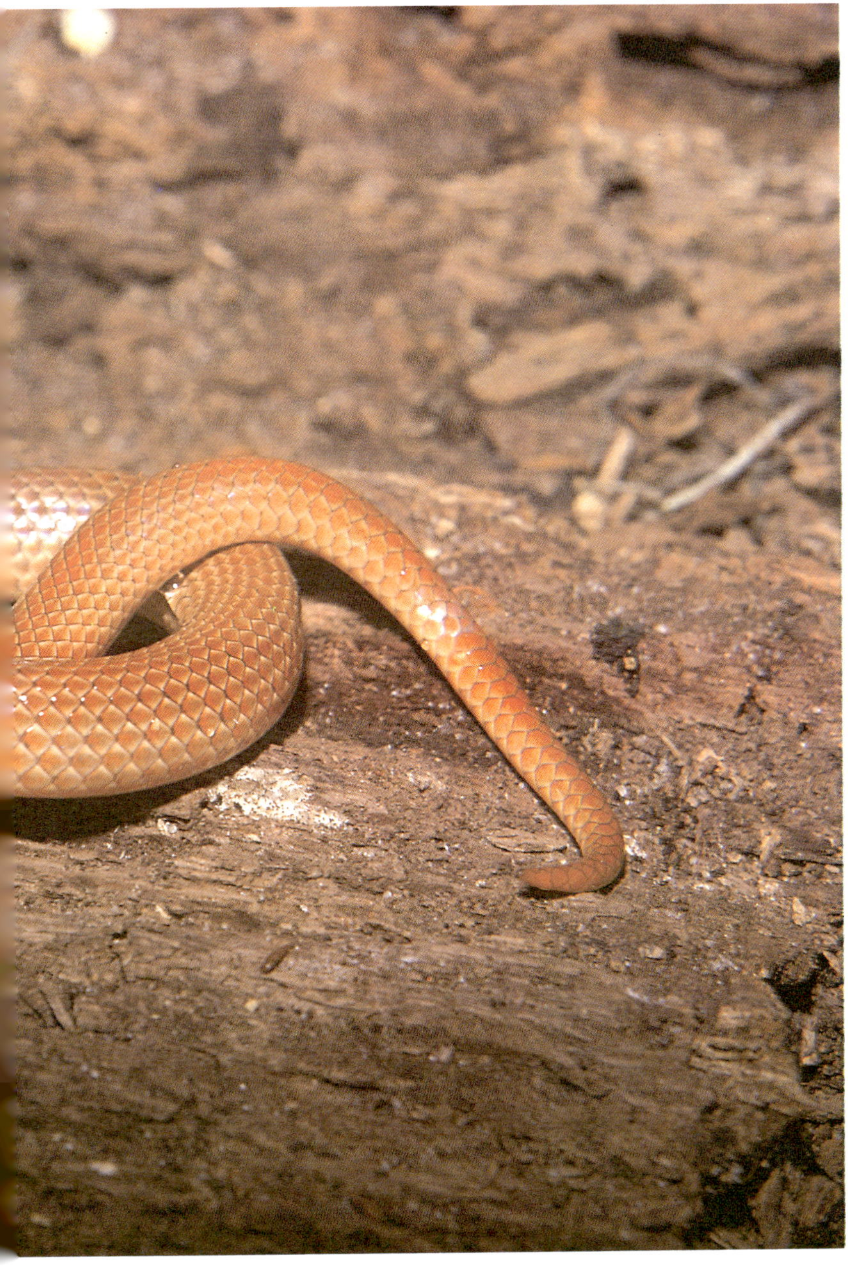

Sonora semiannulata, ✱ 🌒

Sonora semiannulata, ✱ 🌒 50cm

Sonora semiannulata, ✳ 🜚 50cm

Sonora semiannulata, ✳ 🜚 50cm

Sonora michoacanensis, ✹ ☂ 50cm

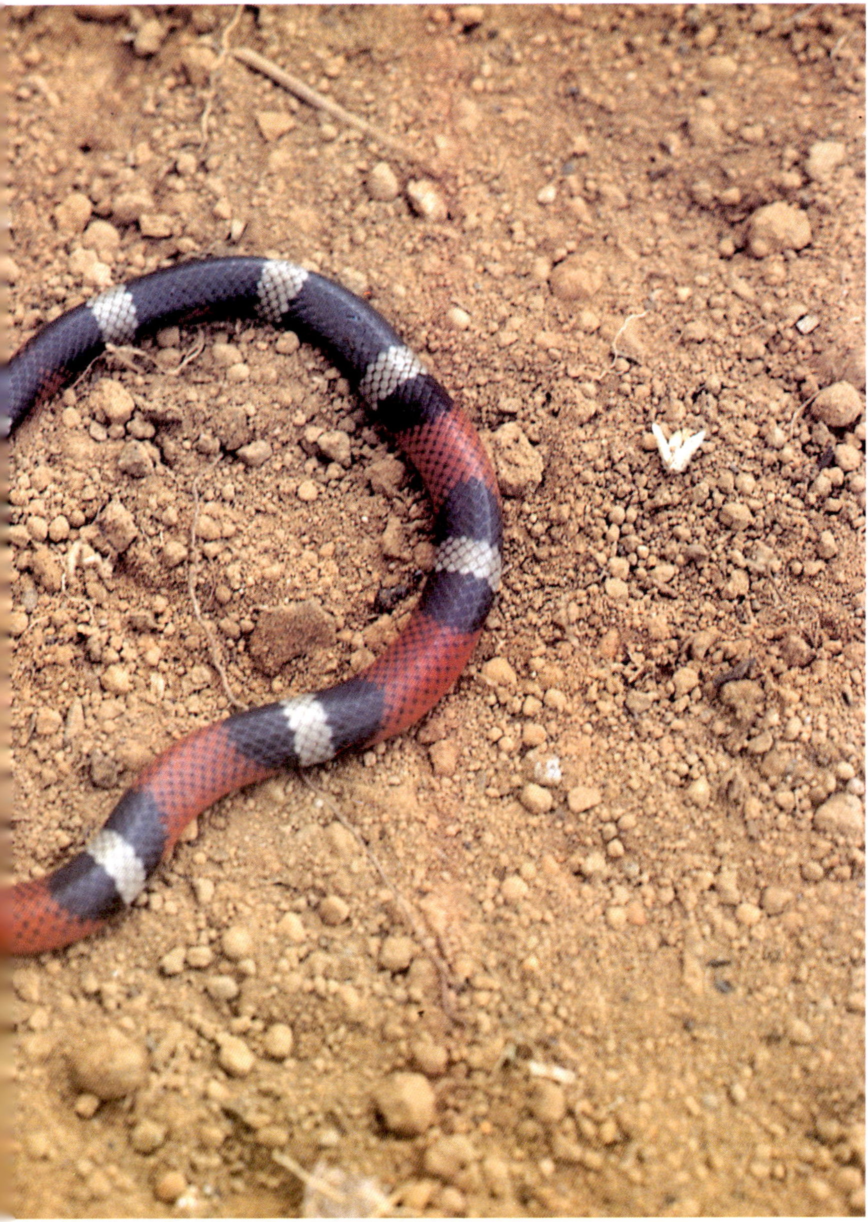

Stenorrhina freminvillei, 🦎 🕷 ⚲ 75cm

Tantilla nigriceps, 🕷 ⚲ 35cm

Tantilla nigriceps, ✴ 🜪 35cm

Tantilla rubra cucullata, ✴ 🜪 35cm

Telescopus dhara, 🐢 🦎 🥚 ☠ 70cm
Telescopus semiannulatus, 🐢 🦎 🥚 ☠ 70cm

Telescopus semiannulatus, 🐍 🦎 ☾ ☠ 70cm

Telescopus semiannulatus, 🐍 🦎 ☾ ☠ 70cm

385

Telescopus beetzii, 🐃 🦎 🥚 ☠ 70cm

Telescopus fallax, 🐃 🦎 🥚 ☠ 75cm

Thamnodynastes strigatus, 80cm

Thelotornis kirtlandi capensis, 130cm

Tomodon dorsatus, 🦎 💧 ☠ 75cm

Tomodon ocellatus, 🦎 💧 ☠ 75cm

Trimorphodon biscutatus vandenburghi, 🐁 🦎 ♂ 120cm
Trimorphodon biscutatus vilkinsoni, 🐁 🦎 ♂ 120cm

Trimorphodon biscutatus lambda, 🐭 🦎 ☾ 120cm
Trimorphodon biscutatus biscutatus, 🐭 🦎 ☾ 120cm

Xenocalamus mechovi inornatus, 🦎 ♂ 50cm
Xenocalamus mechovi inorntaus, 🦎 ♂ 50cm

Dipsas sp., ✱ 🐌 90cm

Dipsas catesbyi, ✱ 🐌 90cm

Sibon nebulata, 🐌 ✳ ⚇ 100cm

Pareas margaritophorus, ✳ ☾ 90cm

Pareas carinatus, ✳ ☾ 90cm

Pareas margaritophorus, 🕷 🐌 90cm

Pareas formosensis, 🕷 🐌 90cm

Aparallactus capensis, ✳ ☾ ☠ 40cm

Aparallactus capensis, ✳ ☾ ☠ 40cm

Chilorhinophis gerardi, ✳ 🐍 ☠ 40cm

Homoroselaps lacteus, 🦎 🐍 50cm

397

Acanthophis antarcticus, 60cm

Acanthophis pyrrhus, 🐍 🦎 🌀 ☠ 60cm
Acanthophis antarcticus, 🐍 🦎 🌀 ☠ 60cm

Aspidelaps lubricus, 🐾 🐸 ☾ ☠ 80cm

Aspidelaps scutatus, 🐾 🐸 ☾ ☠ 80cm

400

Aspidelaps lubricus, 🦎 🐸 🥚 ☠ 80cm
Aspidelaps scutatus, 🦎 🐸 🥚 ☠ 80cm

Austrelaps superbus, 🐸 🐁 ☠ 170cm

Austrelaps superbus, 🐸 🐁 ☠ 170cm

402

Boulengerina annulata, 🐟 💧 ☠ 120cm
Bungarus fasciatus, 🐟 🦎 💧 ☠ 200cm

Bungarus multicinctus, 200cm

Bungarus caeruleus, 🐟 🦎 🥚 ☠ 200cm
Bungarus multicinctus, 🐟 🦎 🥚 ☠ 200cm

Bungarus fasciatus, 🐟 🦎 ☃ ☠ 200cm

Bungarus caeruleus, 🐟 🦎 ☃ ☠ 200cm

Bungarus candidus, 🐟 🦎 🐁 ☠ 200cm
Bungarus flaviceps, 🐟 🦎 🐁 ☠ 200cm

407

Bungarus candidus, 🐟 🦎 🥚 ☠ 200cm
Cacophis squamulosus, 🦎 🕷 🐍 90cm

Calliophis sauteri, 🐾 ☂ ☠ 50cm
　　　　　　　　Calliophis macclellandi, 🐾 ☂ ☠ 50cm

Cryptophis nigrescens, 🦎 🐸 🐍 ☠ 80cm
Cryptophis nigrescens, 🦎 🐸 🐍 ☠ 80cm

Demansia psammophis, 🦎🐸♻☠ 85cm
Dendroaspis jamesoni, 🐁🐦♻☠ 200cm

Dendroaspis angusticeps, 🐢 🐦 ☂ ☠ 220cm
Dendroaspis polylepis, 🐢 🐦 ☂ ☠ 250cm

Denisonia punctata, 🐢 🦎 🐍 ☠ 50cm

Drysdalia coronoides, 🦎 🕷 🐍 ☠ 40cm

Elapsoidea sundevalli boulengeri, 🐾 ● ✧ ☠ 100cm
Furina diadema, 🐾 ✳ ✧ ☠ 40cm

414

Hemachatus haemachatus, 🐸 🦎 🐍 ☠ 120cm
Hemachatus haemachatus, 🐸 🦎 🐍 ☠ 120cm

Hemachatus haemachatus, 🐢 🦎 🐸 ☠ 120cm
Hemiaspis signata, 🦎 🐸 🐍 ☠ 50cm

Hoplocephalus bitorquatus, 🦎 🐸 🥚 ☠ 65cm
Micruroides euryxanthus, 🦎 🕷 🥚 ☠ 55cm

Micrurus fulvius tenere, 🦎 🌀 ☠ 75cm

Micrurus fulvius, 🦎 🌀 ☠ 75cm

Micrurus nigrocinctus, 🦎 ☝ ☠ 120cm

Micrurus dumerili, 🦎 ☝ ☠ 100cm

Micrurus fulvius barbouri, 🦎 🌀 ☠ 75cm

Micrurus fulvius, 🦎 🌀 ☠ 75cm

Micrurus corallinus corallinus, 🦎 🌙 ☠ 125cm
Micrurus frontalis frontalis, 🦎 🌙 ☠ 120cm

Micrurus fulvius, 🦎 ☂ ☠ 75cm

423

Micrurus fulvius tenere, 🦎 ☾ ☠ 75cm

Micrurus lemniscatus, 🦎 ☾ ☠ 105cm

Micrurus nigrocinctus nigrocinctus, 95cm

Micrurus fulvius, 75cm

Naja naja, 🐭 🦎 🥚 ☠ 200cm

Naja nigricollis, 220cm

Naja naja kaouthia, 🐀 🦎 ☁ ☠ 200cm

Naja naja kaouthia, 🐀 🦎 ☁ ☠ 200cm

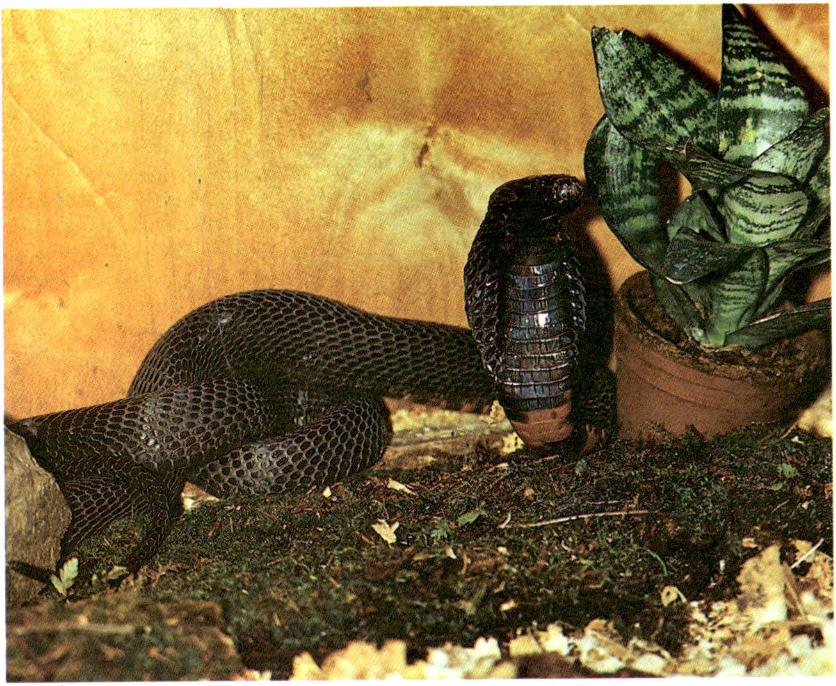

Naja nigricollis, 220cm

Naja oxiana, 190cm

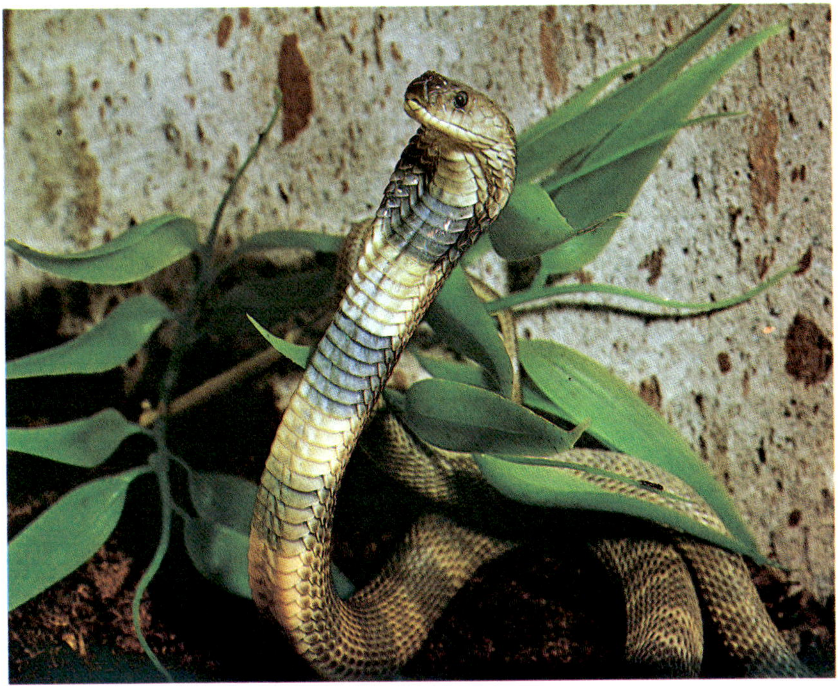

Naja naja philippinensis, 🐟 🦎 ☙ ☠ 200cm

Naja naja sputatrix, 🐟 🦎 ☙ ☠ 200cm

Naja mossambica pallida, 🐁 🦎 ☁ ☠ 120cm
Naja naja sumatrana, 🐁 🦎 ☁ ☠ 200cm

Naja melanoleuca, 🐸 ✋ ☠ 200cm

Naja naja, 🐁 🦎 ✋ ☠ 200cm

432

Naja mossambica, 🐀 🦎 🐍 ☠ 120cm
Naja nivea, 🐀 🦎 🐍 ☠ 150cm

Naja melanoleuca, 🐸 🦆 ☘ 200cm

Naja haje, 200cm

Notechis scutatus, 150cm

435

Notechis scutatus, 🐀 🦎 🐸 ☠ 150cm
Notechis scutatus, 🐀 🦎 🐸 ☠ 150cm

Ophiophagus hannah, 420cm

Ophiophagus hannah, 420cm

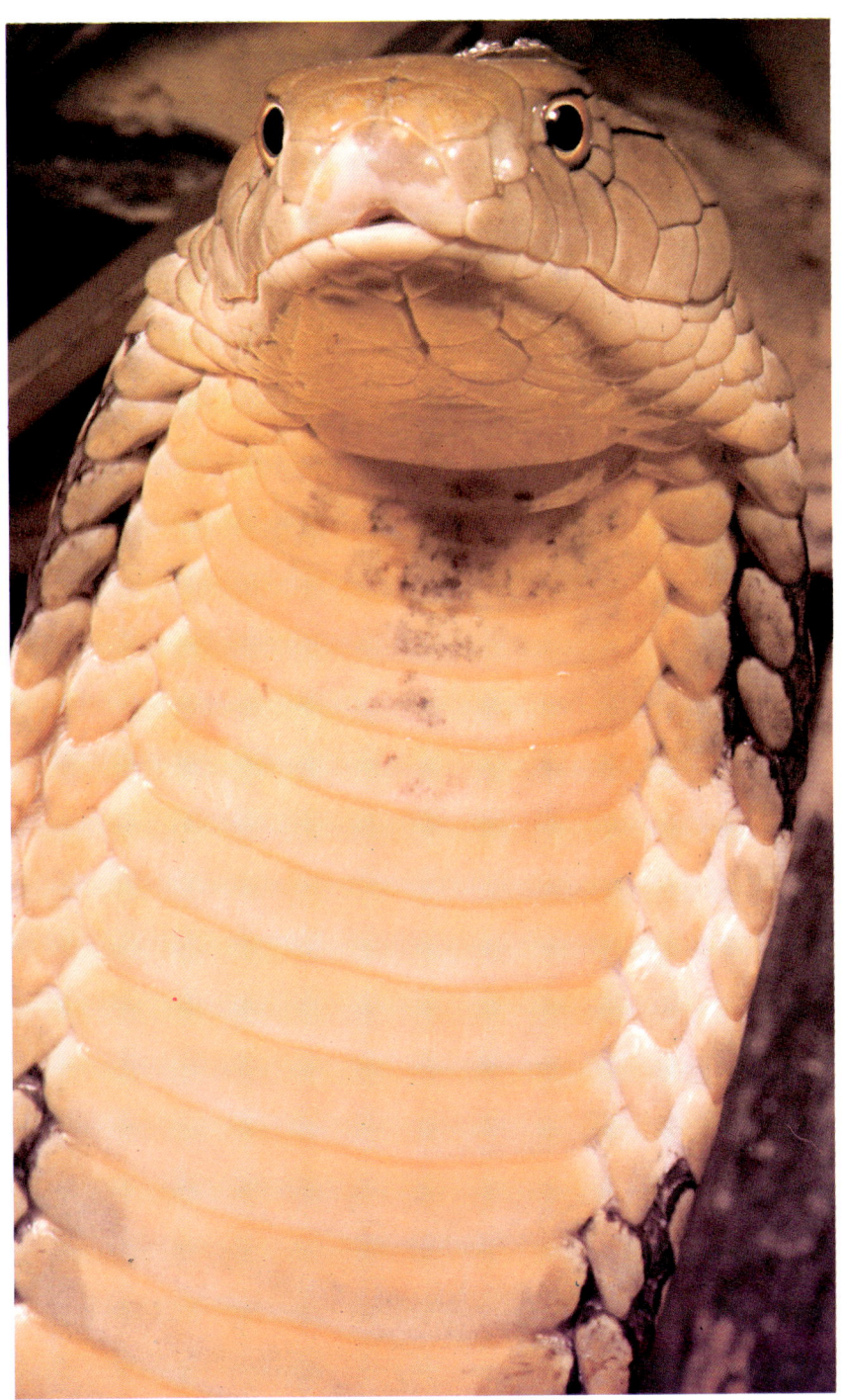

Ophiophagus hannah, 🐍 ☁ ☠ 420cm

Ophiophagus hannah, 🐾 ☾ ☠ 420cm

Oxyuranus scutellatus, 🐁 ☾ ☠ 300cm

Oxyuranus scutellatus, 🐀 ☁ ☠ 300cm
Pseudechis guttatus, 🐀 🦎 ☠ 190cm

Pseudechis australis, 🐀 🦎   ✼ 190cm

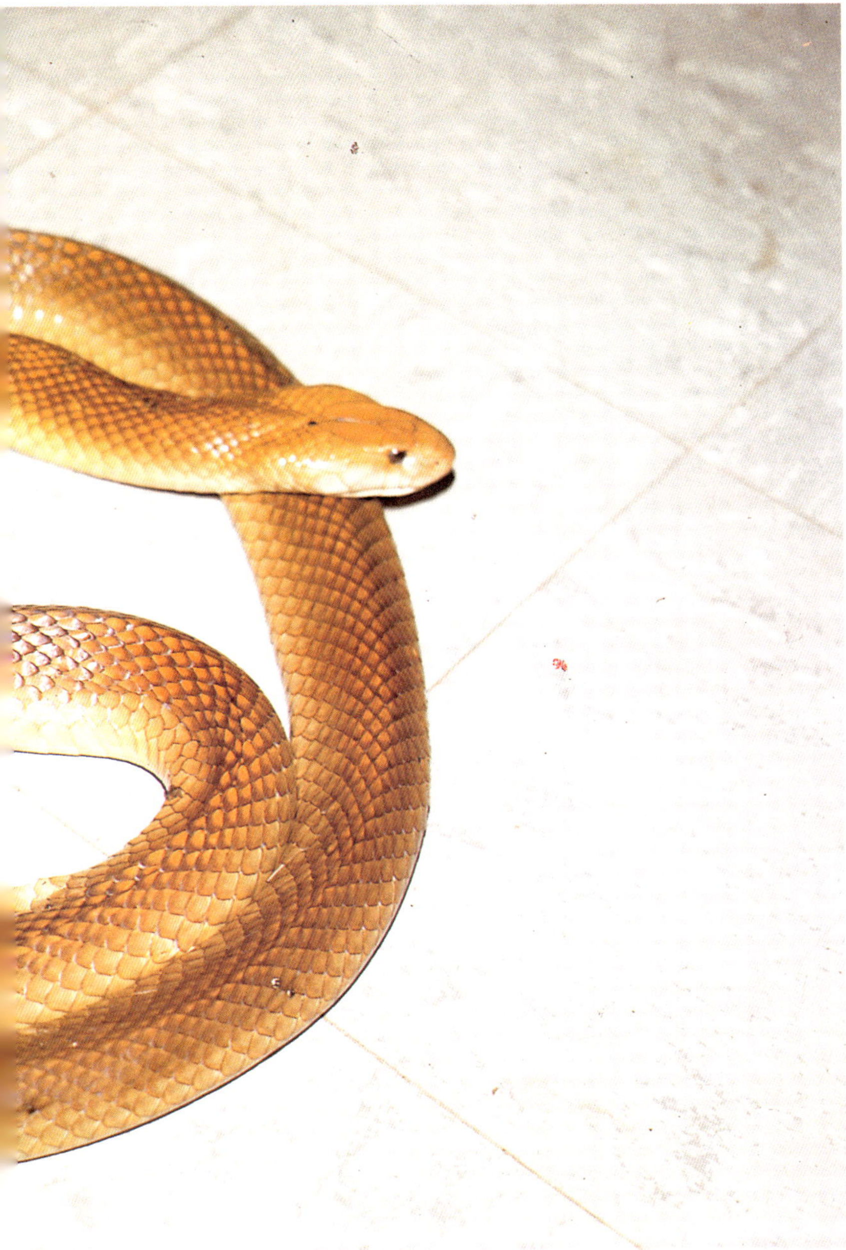

Pseudechis porphyriacus, 🐾 🦎 🐸 ☠ 190cm

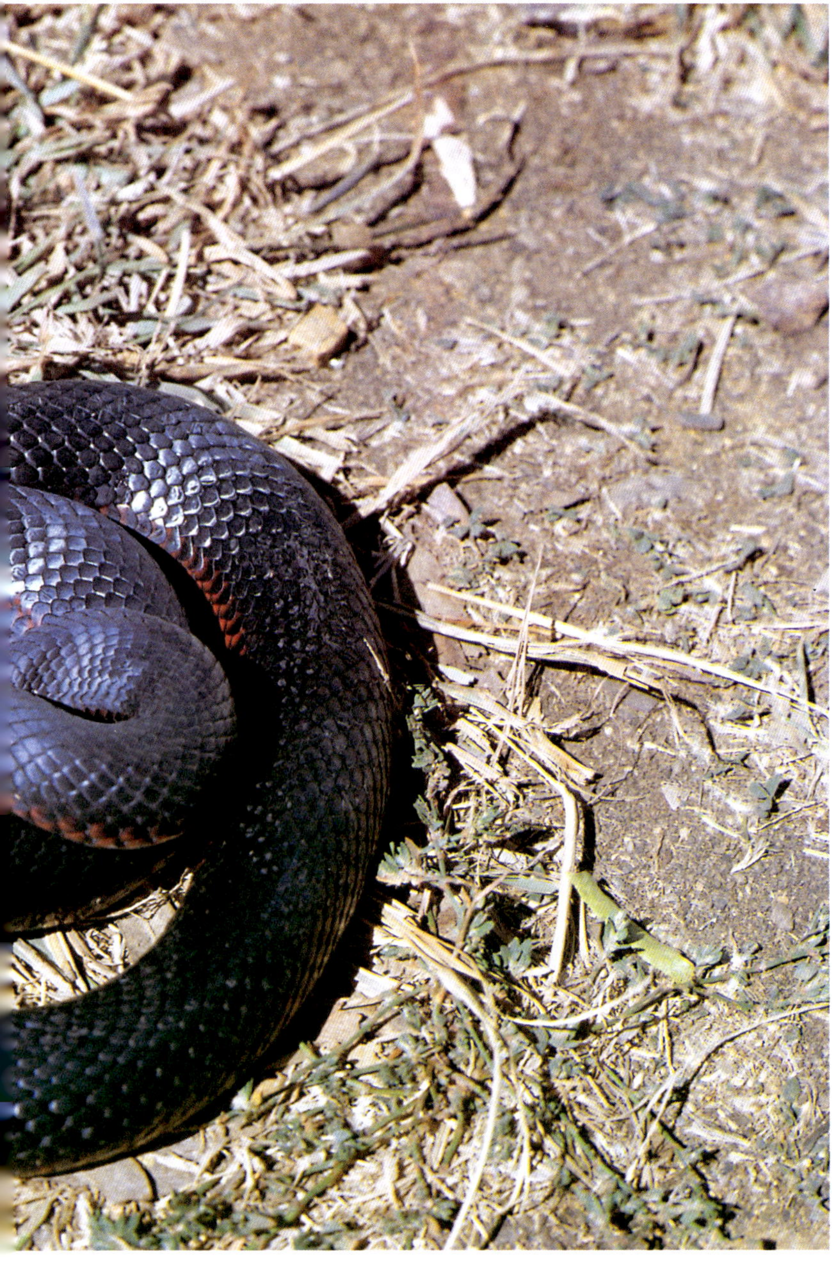

Pseudonaja textilis, 🐀 🦎 🥚 ☠ 105cm

Suta suta, 🦎 🪱 ☠ 30cm

Unechis nigrostriatus, 🦎 🐍 ☠ 50cm
　　　　　　　　　　Unechis brevicaudus, 🦎 🐍 ☠ 50cm

Unechis gouldi, 🦎 🌀 ☠ 50cm

Vermicella annulata, 🦎 🐌 ☠ 50cm

Vermicella annulata, 🦎 🐚 ☠ 50cm
Walterinnesia aegyptia, 🦎 🐸 🐚 ☠ 110cm

Walterinnesia aegyptia, 🦎 🐸 🥚 ☠ 110cm
Walterinnesia aegyptia, 🦎 🐸 🥚 ☠ 110cm

Laticauda sp., 🐟 🐚 ☠ 110cm

Laticauda colubrina, 🐟 🐚 ☠ 110cm

451

Laticauda semifasciata, 🐟 🐚 ☠ 110cm

Laticauda sp., 🐟 🐚 ☠ 110cm

Laticauda colubrina, 🐟 🌙 ☠ 110cm

Acalyptophis peroni, 🐟 🌙 ☠ 100cm

Aipysurus apraefrontalis, 🐟 🌀 ☠ 80cm

Aipysurus fuscus, 🐟 🌀 ☠ 80cm

Aipysurus sp., 🐟 💀 80cm

Disteira [Astrotia] stokesi, 🐟 🕷 💀 90cm

Disteira [Astrotia] stokesi, 🐟 ✳ 🌙 ☠ 90cm

Emydocephalus annulatus, 🐟 🌀 ☠ 75cm

Enhydrina schistosa, 🐟 🌀 ☠ 90cm

Hydrophis cyanocinctus, 🐟 🌀 ☠ 90cm

Hydrophis cyanocinctus, 🐟 〰️ ☠ 90cm

Hydrophis semperi, 🐟 〰️ ☠ 90cm

Hydrophis inornatus, 🐟 🌀 ☠ 90cm

Pelamis platurus, 🐟 🌀 ☠ 65cm

Pelamis platurus, 🐟 ☠ 65cm

Pelamis platurus, 🐟 ☠ 65cm

Azemiops feae, 🐁 ☾ ☠ 90cm

Azemiops feae, 🐁 ☾ ☠ 90cm

Atheris squamiger, 🐢 🦎 🌙 ☠ 75cm

Atheris squamiger, 75cm

Atheris squamiger, 75cm

Atheris squamiger, 75cm

Atheris squamiger, 🐍 🦎 🌙 ☠ 75cm

Atheris nitschei, 🐾 🦎 🌀 ☠ 75cm

Atheris hispidus, 🐾 🦎 🌀 ☠ 75cm

Atheris chloroechis, 🐾 🦎 🌀 ☠ 75cm

Atheris hispidus, 🐾 🦎 🌀 ☠ 75cm

469

Atheris superciliaris, 🐟 🦎 🐁 ☠ 75cm

Atheris squamiger, 🐟 🦎 🐁 ☠ 75cm

Bitis caudalis, 75cm

Bitis caudalis, 75cm

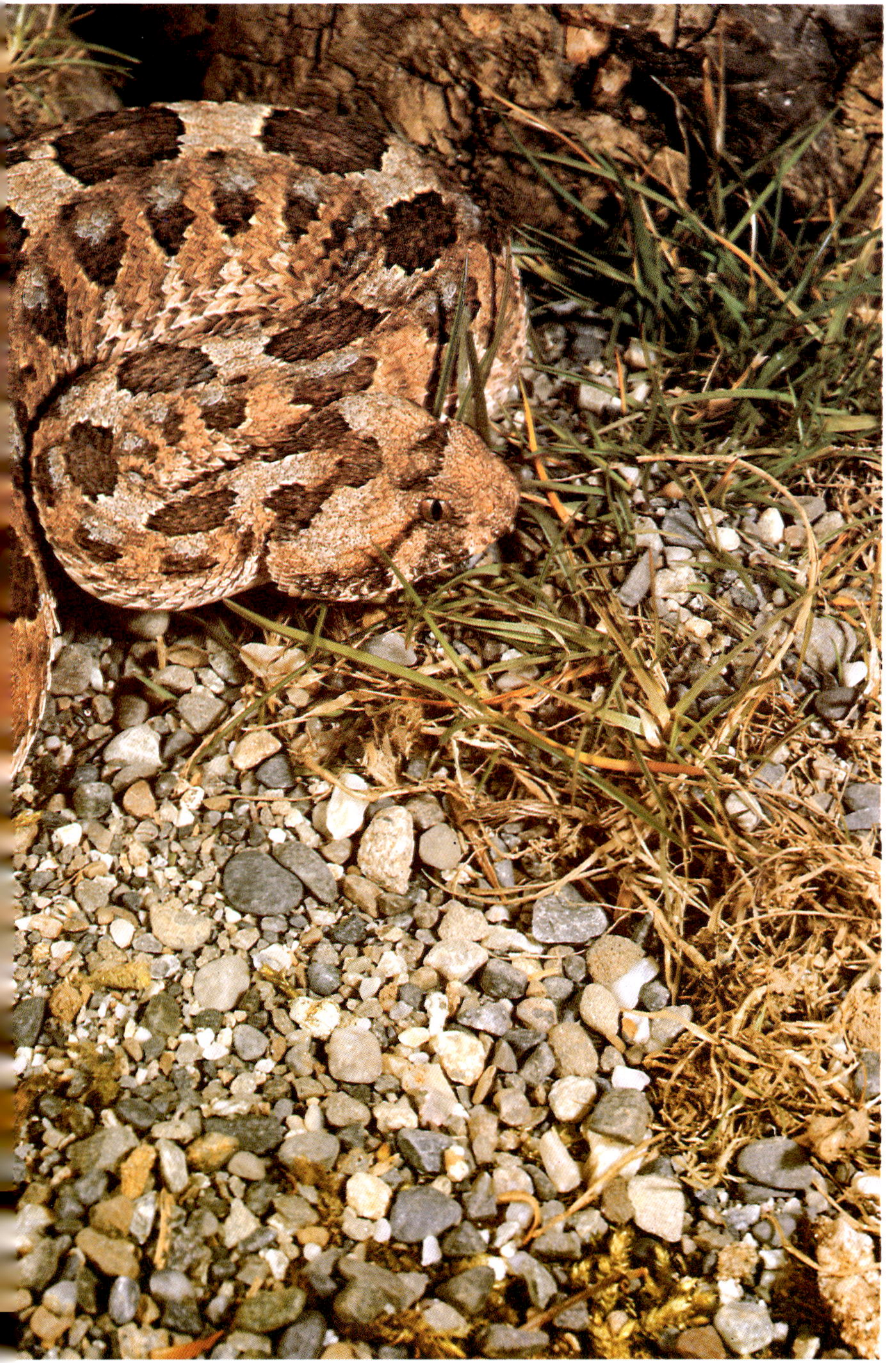

Bitis caudalis, 75cm

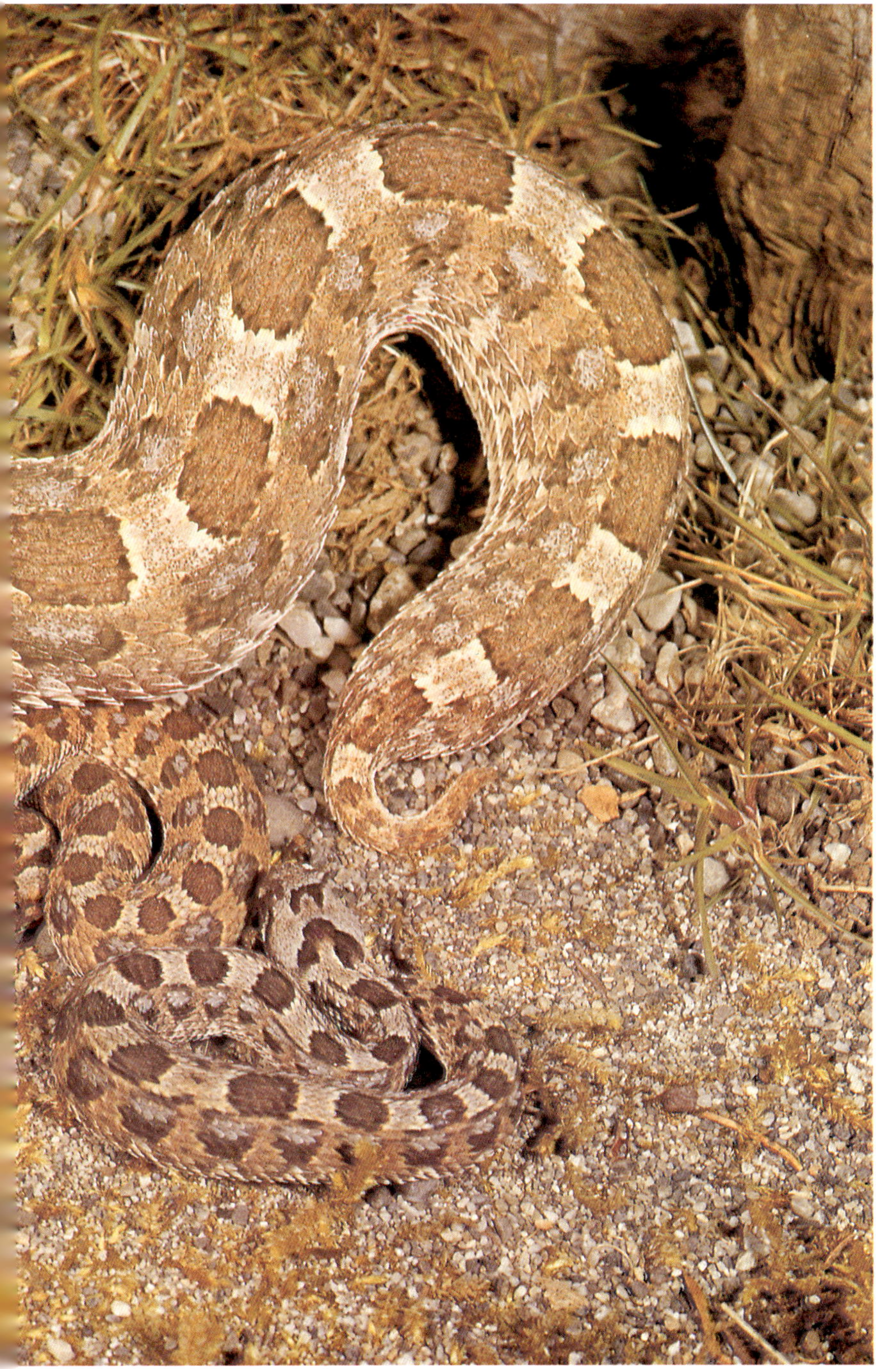

Bitis nasicornis, 🐟 🐍 ☠ 170cm
Bitis gabonica rhinoceros, 🐟 🐦 🐍 ☠ 120cm

476

Bitis arietans, 🐀 🐦 🐸 ☠ 150cm

Bitis gabonica, 🐀 🐦 🐸 ☠ 120cm

Bitis nasicornis, 🐀 🌀 ☠ 170cm

Bitis arietans, 🐀 🐦 🌀 ☠ 150cm

Bitis atropos, 🐀 🦎 🌀 ☠ 60cm

Bitis gabonica, 🐀 🐦 🌀 ☠ 120cm

479

Bitis inornata, 🐢 🦎 🌀 ☠ 50cm

Bitis arietans, 🐢 🐦 🌀 ☠ 150cm

Bitis caudalis, 🐢 🦎 🌀 ☠ 75cm

Bitis peringueyi, 🦎 🌀 ☠ 30cm

Bitis caudalis, 75cm

Bitis xeropaga, 🐢 🦎 💀 50cm

Bitis peringueyi, 🦎 💀 30cm

Bitis schneideri, 🦎 🌀 ☠ 20cm

Bitis cornuta, 🐁 🦎 🌀 ☠ 60cm

484

Cerastes vipera, 🐁 🦎 🌙 ☠ 75cm

Cerastes cerastes, 75cm

Echis coloratus, 80cm

Echis carinatus, 🐀 🦎 🥚 ☠ 80cm

Echis coloratus, 🐀 🦎 🥚 ☠ 80cm

Echis carinatus, 🐁 🦎 🥚 ☠ 80cm

Echis coloratus, 🐁 🦎 🥚 ☠ 80cm

Echis carinatus, 🐁 🦎 🥚 ☠ 80cm

Eristicophis macmahonii, 60cm

Vipera [Daboia] russelli, 🐀 🐦 🌀 ☠ 135cm

Vipera [Pseudocerastes] fieldi, 🐀 🦎 🌀 ☠ 100cm

Vipera aspis, 🐁 🐦 🦎 ☠ 40cm

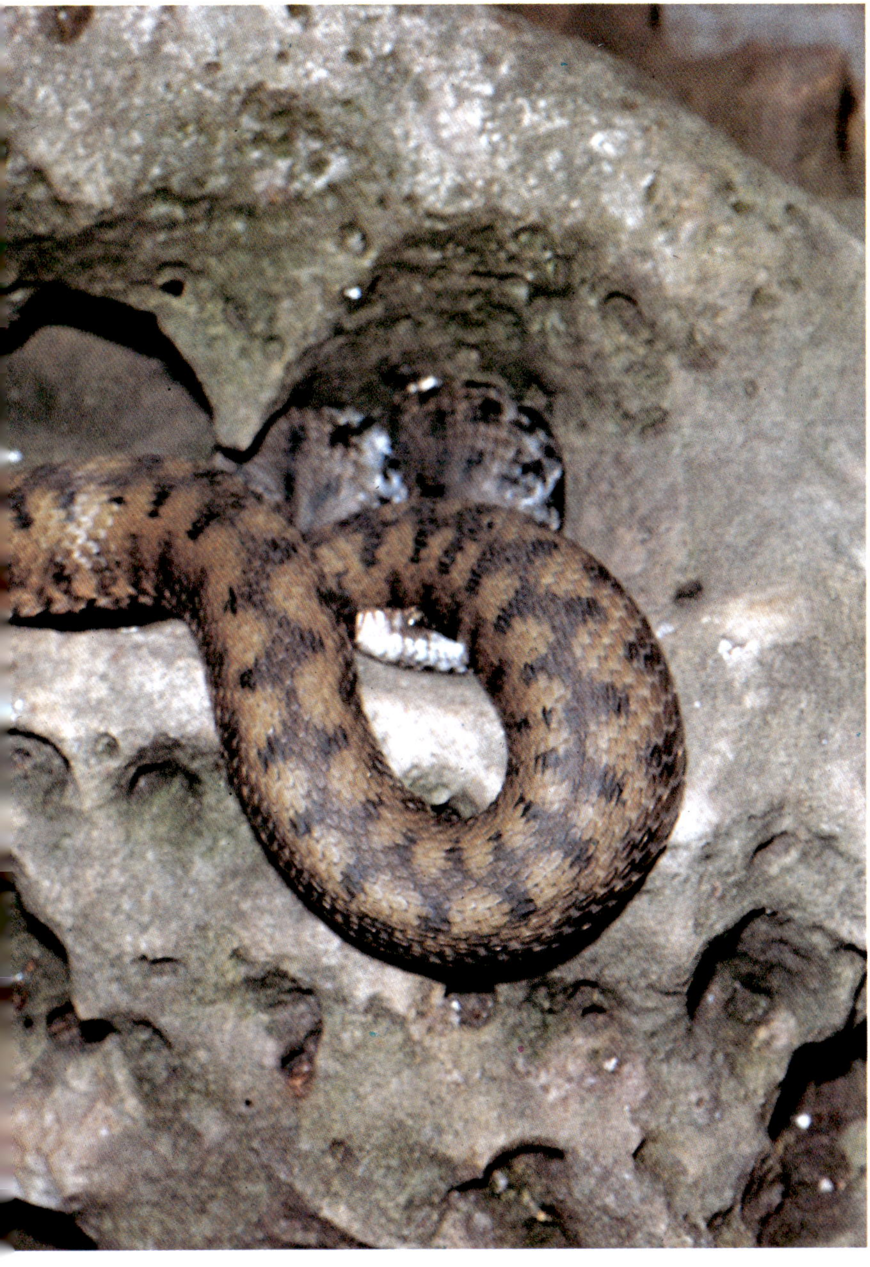

Vipera xanthina, 100cm

Vipera lebetina, 🐀 ⚡ / 🍌 ☠ 75cm

Vipera xanthina, 100cm

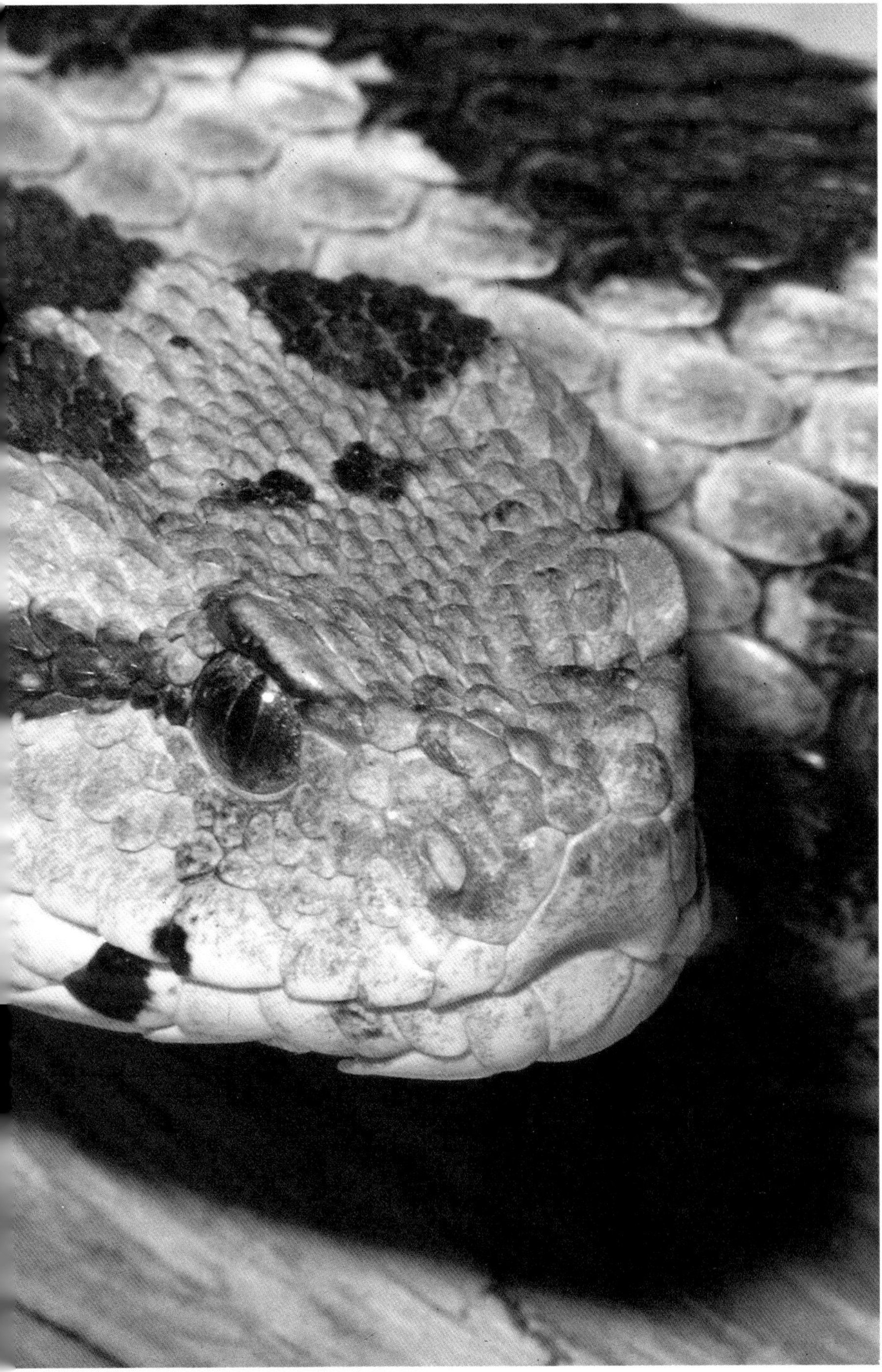

Vipera berus, 🐸 🦎 🍌 ☠ 60cm

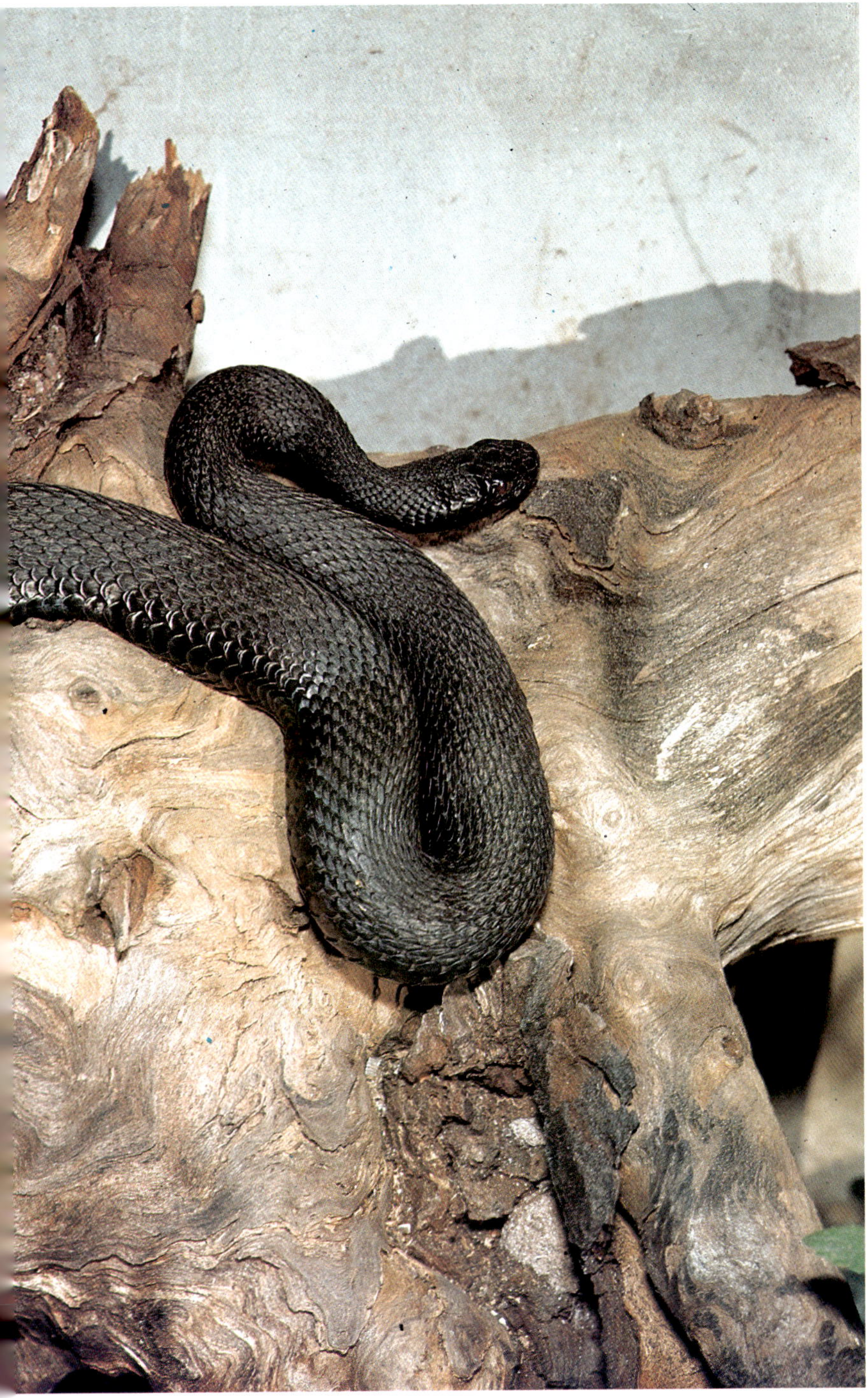

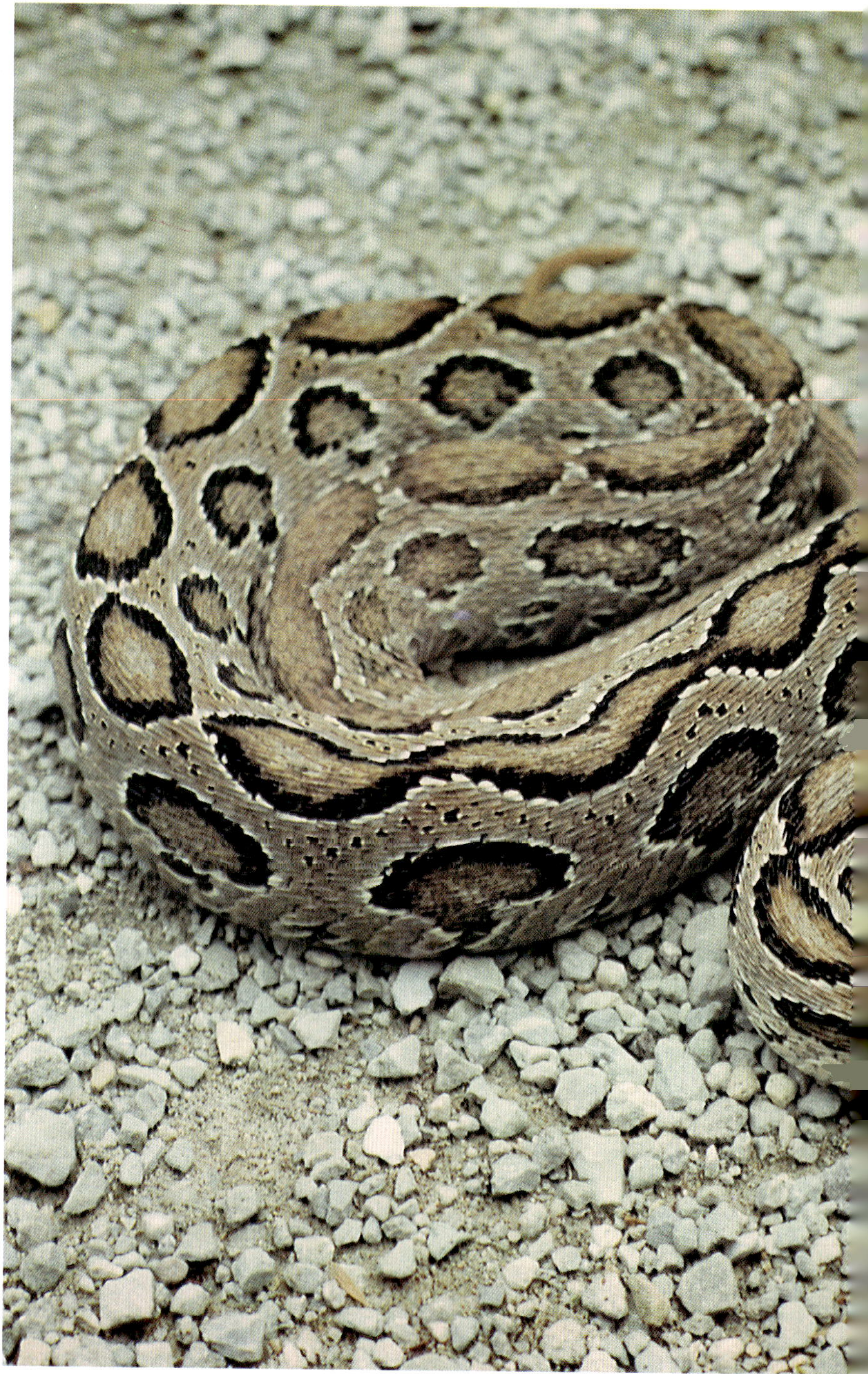

Vipera [Daboia] russelli, 135cm

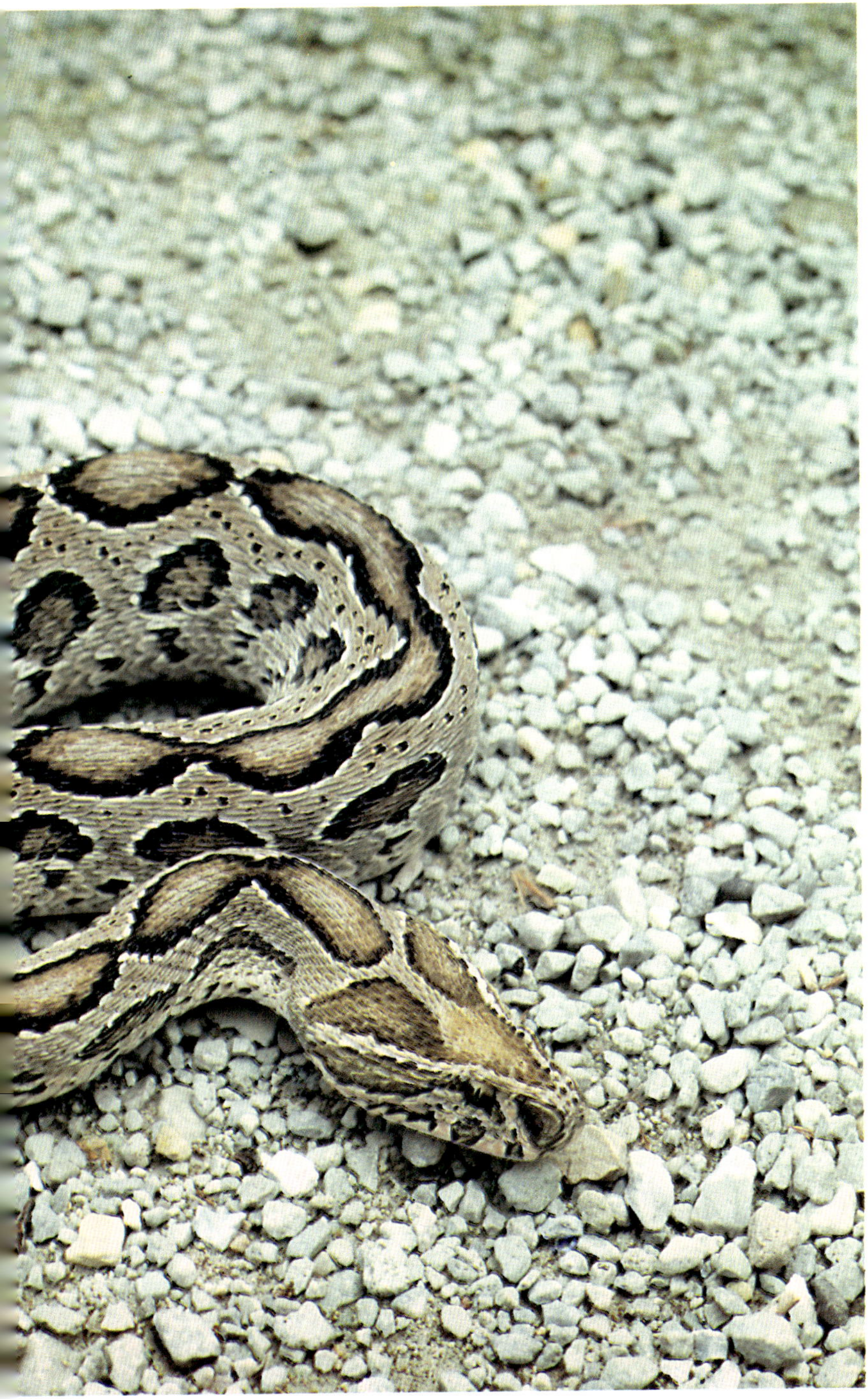

Vipera palaestinae, 🐁 🐦 ✋ ☠ 100cm

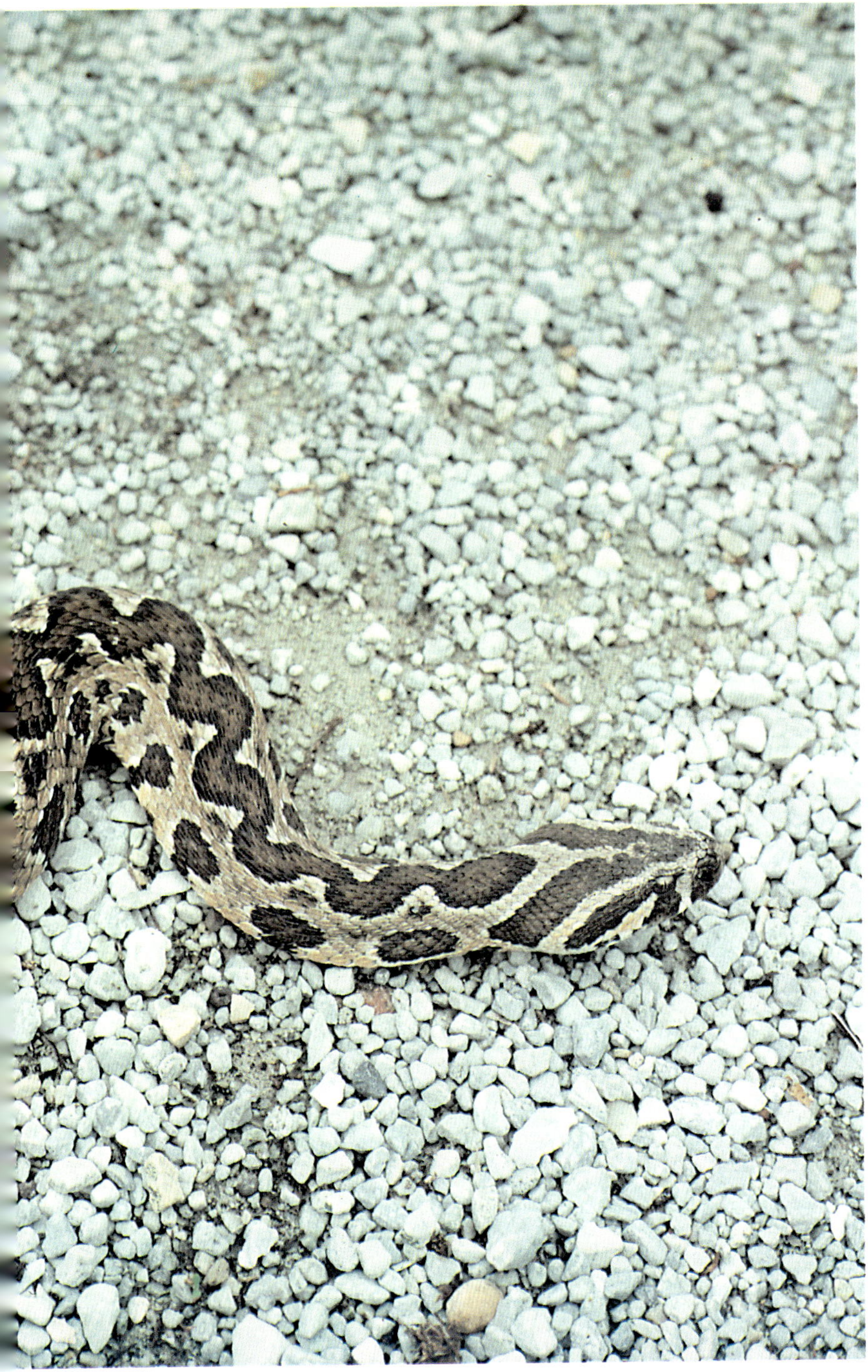

Vipera berus, 🐸 🦎 🌀 ☠ 60cm

Vipera berus, 🐸 🦎 🌀 ☠ 60cm

Vipera berus, 🐢 🦎 🌀 ☠ 60cm

Vipera kaznakovi, 🐢 🦎 🌀 ☠ 60cm

Vipera bornmulleri, 🐢🐦🐍☠100cm

Vipera kaznakovi, 🐢🦎🐍☠60cm

Vipera lebetina, 🐢 ♋ / 🐍 ☠ 75cm

Vipera latifi, 🐢 🐦 ♋ ☠ 100cm

Vipera berus, 🐢 🦎 🌀 ☠ 60cm

Vipera [Daboia] russelli, 🐢 🐦 🌀 ☠ 135cm

Vipera raddei, 🐢 🐦 🌀 ☠ 95cm
Vipera ammodytes transcaucasiana, 🐢 🐦 🌀 ☠ 70cm

Causus rhombeatus, 65cm

Causus defilippi, 🐸 ☁ ☠ 65cm

Causus maculatus, 🐸 ☁ ☠ 65cm

Causus rhombeatus, 🐸 ☂ ☠ 65cm *Causus rhombeatus,* 🐸 ☂ ☠ 65cm

512

Agkistrodon bilineatus taylori, 🐟 🐠 〰 ☠ 90cm
Agkistrodon bilineatus, 🐟 🐠 〰 ☠ 90cm

Agkistrodon bilineatus, 90cm

Agkistrodon [Deinagkistrodon] acutus, 🐟 🐸 🥚 ☠ 120cm
Agkistrodon piscivorus, 🐟 🐟 🥚 ☠ 120cm

Agkistrodon piscivorus leucostoma, 120cm

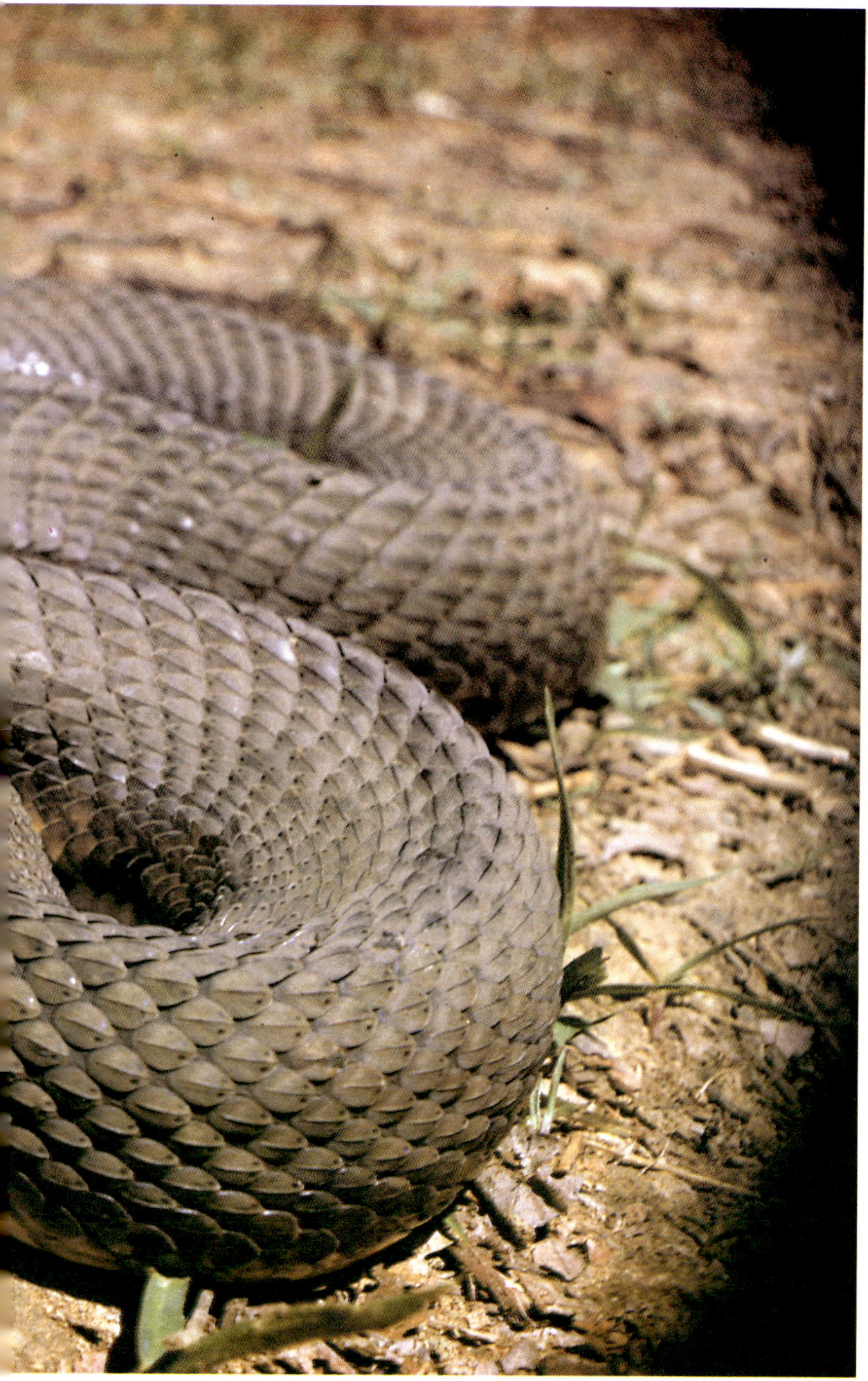

Agkistrodon bilineatus, 90cm

Agkistrodon bilineatus, 🐀 🐟 🍌 ☠ 90cm

Agkistrodon piscivorus, 🐟 🐠 〰️ ☠️ 120cm
Agkistrodon piscivorus, 🐟 🐠 〰️ ☠️ 120cm

Agkistrodon bilineatus taylori, 🐟 🐸 🦎 ☠ 90cm
Agkistrodon contortrix, 🐟 🐸 🦎 ☠ 80cm

Agkistrodon contortrix mokasen, 🐟 🐸 🦎 ☠ 80cm
Agkistrodon contortrix mokasen, 🐟 🐸 🦎 ☠ 80cm

Agkistrodon piscivorus, 🐟 🐸 🌀 ☠ 120cm
Agkistrodon contortrix mokasen, 🐭 🐸 🌀 ☠ 80cm

Agkistrodon contortrix mokasen, 🐭 🐸 🐍 ☠ 80cm
Agkistrodon bilineatus, 🐭 🐟 🐍 ☠ 90cm

524

Agkistrodon bilineatus russeolus, 🐟 🐠 🦎 ☠ 90cm
Agkistrodon contortrix laticinctus, 🐁 🐸 🦎 ☠ 80cm

Agkistrodon bilineatus, 🐁 🐸 🐟 ☠ 90cm
Agkistrodon blomhoffi brevicaudis, 🐁 🐦 🐟 ☠ 75cm

Agkistrodon blomhoffi brevicaudis, 🐀 🐦 🐸 ☠ 75cm

Agkistrodon intermedius caucasicus, 🐁 🦎 🌀 ☠ 100cm
Agkistrodon bilineatus howardgloydi, 🐁 🐟 🌀 ☠ 90cm

Agkistrodon [Hypnale] hypnale, 🐟 🐸 🦎 ☠ 85cm

Agkistrodon rhodostoma, 🐟 🐸 🦎   75cm

Agkistrodon [Hypnale] hypnale, 🐟🐸 🪱   85cm
Agkistrodon [Deinagkistrodon] acutus, 🐟🐸 🐁☠ 120cm

530

Bothrops [Bothriechis] bilineatus, 75cm

Bothrops schlegeli, 75cm

Bothrops [Bothriechis] aurifer, 🐁 🐦 🐸 ☠ 60cm
Bothrops schlegeli, 🐁 🐦 🐸 ☠ 75cm

Bothrops alternatus, 🐢 🦎 ☠ 135cm
Bothrops ammodytoides, 🐢 🦎 ☠ 70cm

Bothrops asper, 🐢 🐸 🌀 ☠ 75cm

Bothrops atrox, 🐢 🌀 ☠ 150cm

Bothrops schlegeli, 🐀 🐦 🐸 ☠ 75cm

Bothrops schlegeli, 🐀 🐦 🐸 ☠ 75cm

Bothrops asper, 🐀 🐦 🐸 ☠ 75cm

Bothrops brazili, 🐢 🐦 🍌 ☠ 85cm
Bothrops nummifer, 🐢 🐦 🍌 ☠ 90cm

Bothrops jararaca, 🐁 🐦 🐍 ☠ 90cm
Bothrops jararaca, 🐁 🐦 🐍 ☠ 90cm

Bothrops ophryomegas, 🐾 🐦 💤 ☠ 90cm

Bothrops schlegeli, 🐾 🐦 🦎 ☠ 75cm

Bothrops rowleyi, 🐀 🐦 🐍 ☠ 80cm

Bothrops nasutus, 🐀 🐦 🐍 ☠ 80cm

Bothrops schlegeli, 🐁 🐦 🐍 ☠ 75cm
Bothrops lateralis, 🐁 🐦 🐍 ☠ 70cm

Bothrops schlegeli, 🐁 🐦 🍃 ☠ 75cm
                        Bothrops neuwiedi, 🐁 🐦 🍃 ☠ 75cm

Bothrops lateralis, 🐀 🐦 🐸 ☠ 70cm

Bothrops schlegeli, 🐀 🐦 🐸 ☠ 75cm

Bothrops jararaca, 🐀 🐦 🐸 ☠ 90cm
Bothrops moojeni, 🐀 🐦 🐸 ☠ 80cm

Bothrops schlegeli, 🐢 🐦 🌿 ☠ 75cm
Bothrops nummifer, 🐢 🐦 🌿 ☠ 90cm

Crotalus adamanteus, 🐢 💨 ☠ 180cm

Crotalus adamanteus, 🐢 💨 ☠ 180cm

Crotalus viridis oreganus, 🐁 🦎 ☠ 110cm

Crotalus molossus, 🐢 🌿 ☠ 105cm
Crotalus mitchelli pyrrhus, 🐢 🌿 ☠ 130cm

Crotalus atrox, 🐟 💨 ☠ 180cm

Crotalus adamanteus, 🐟 💨 ☠ 180cm

Crotalus atrox, 🐁 💤 ☠ 180cm

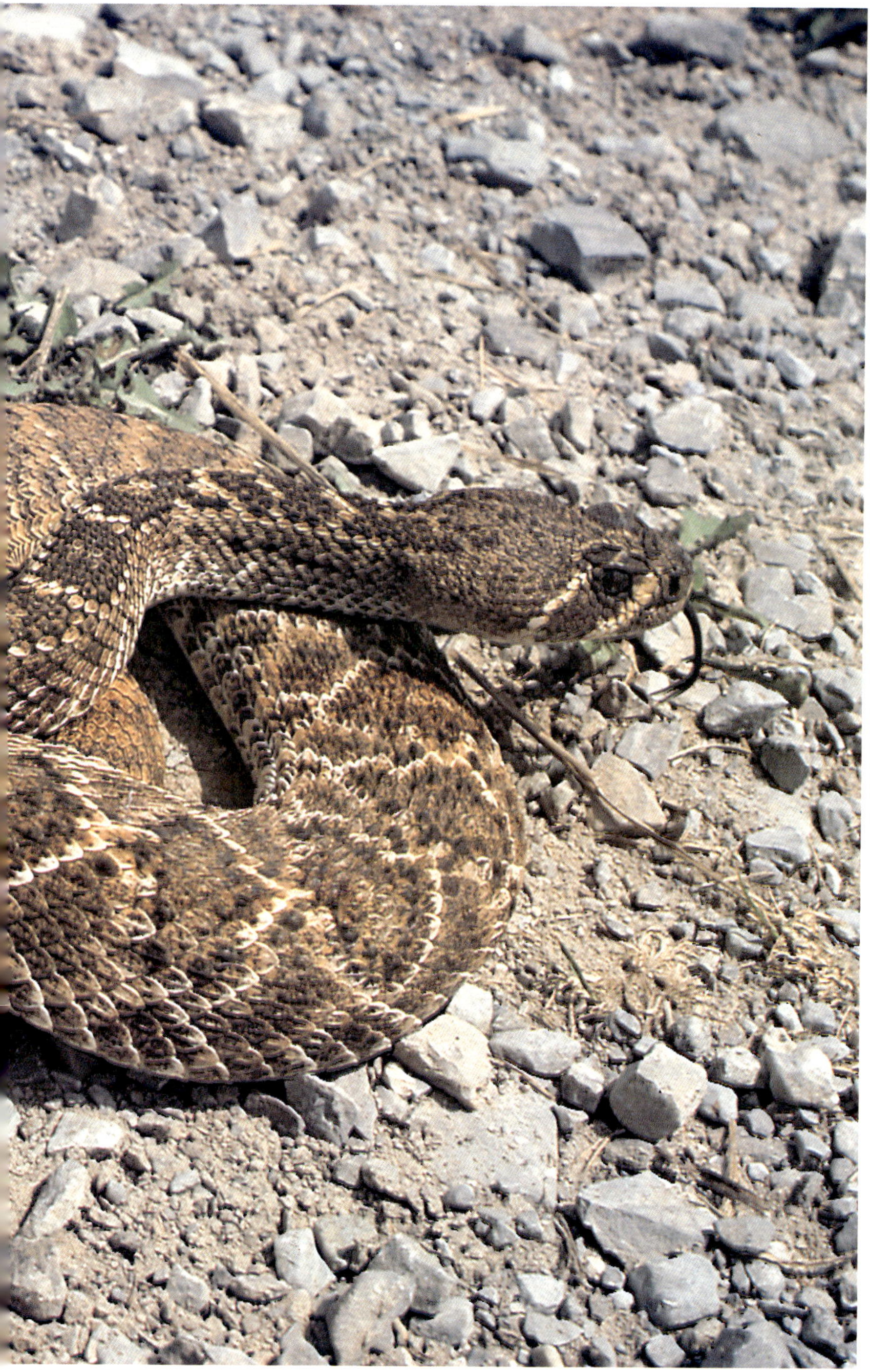

Crotalus ruber, 160cm

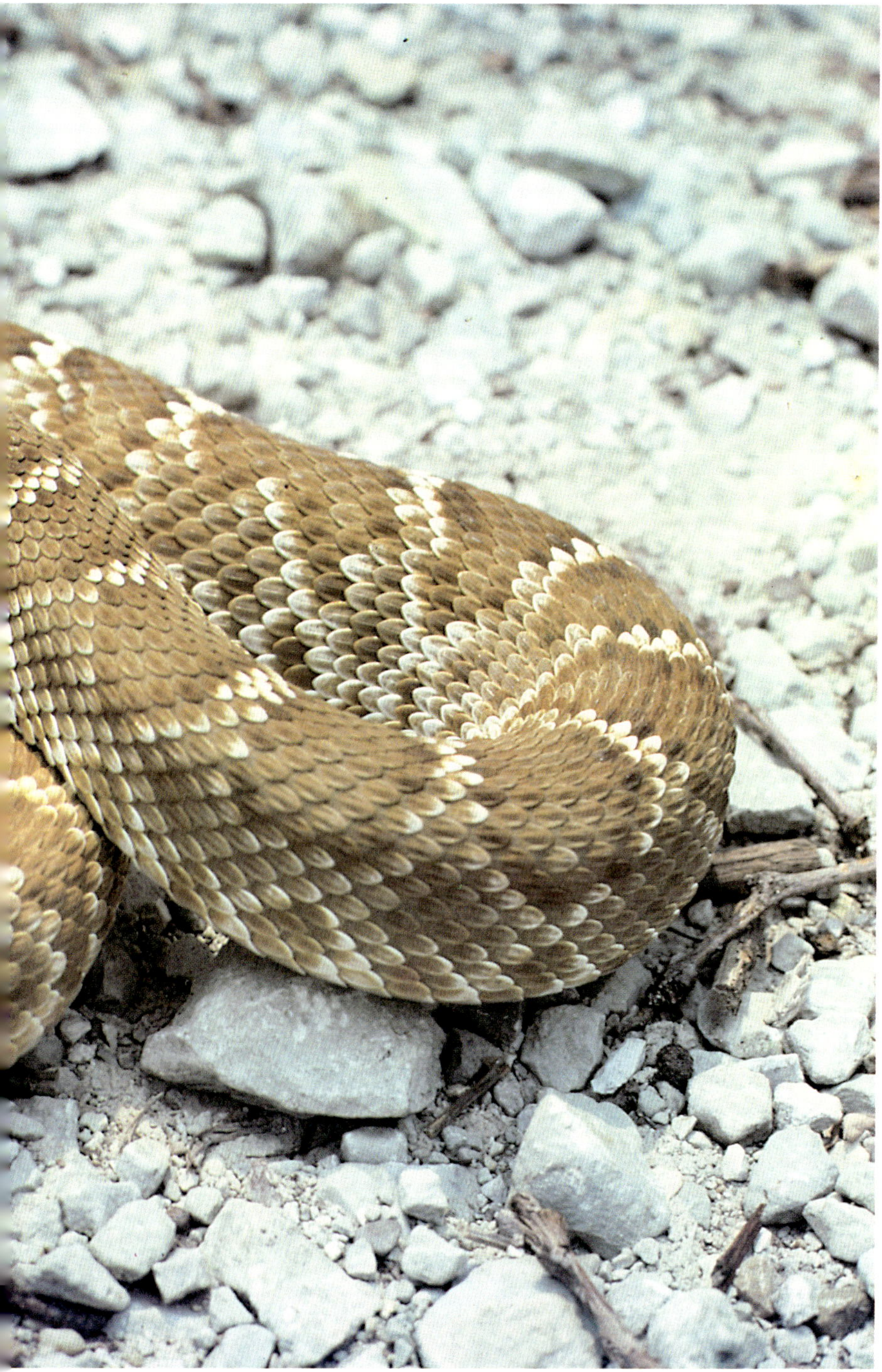

Crotalus atrox, 🐁 🌀 ☠ 180cm

Crotalus lepidus klauberi, 🐁 🌀 ☠ 60cm

Crotalus viridis oreganus, 🐟 🐁 ☠ 110cm
Crotalus horridus atricaudatus, 🐟 🐁 ☠ 150cm

Crotalus adamanteus, 🐟 💨 ☠ 180cm
Crotalus cerastes laterorepens, 🐢 💨 ☠ 80cm

558

Crotalus horridus horridus, 150cm

Lachesis muta, 225cm

Lachesis muta, 225cm

Sistrurus catenatus, 🐢 🐦 🐍 ☠ 75cm
Sistrurus catenatus, 🐢 🐦 🐍 ☠ 75cm

Sistrurus catenatus catenatus, 🐭 🐦 〰 ☠ 75cm
Sistrurus miliarius miliarius, 🐭 🦎 〰 ☠ 50cm

Sistrurus miliarius, 50cm

Trimeresurus stejnegeri, 🐟 🐸 🐍 ☠ 60cm
Trimeresurus purpureomaculatus, 🐀 🐦 🐍 ☠ 80cm

Trimeresurus purpureomaculatus, 🐢 🐦 🐍 ☠ 80cm

Trimeresurus macgregori, 🐢 🌙 ☠ 80cm

Trimeresurus gramineus, 🐢 🌙 ☠ 80cm

Trimeresurus popeorum, 🐀 🦎 💧 ☠ 90cm
Trimeresurus wagleri, 🐀 🐦 💧 ☠ 80cm

Trimeresurus purpureomaculatus, 🐢 🐦 🌀 ☠ 80cm
Trimeresurus popeorum, 🐢 🦎 🌀 ☠ 90cm

Trimeresurus wagleri, 🐁 🐦 🐸 ☠ 80cm
Trimeresurus albolabris, 🐁 🦎 🐸 ☠ 85cm

Trimeresurus wagleri, 🐭 🐦 🦎 ☠ 80cm

Trimeresurus puniceus, 🐭 🐸 🦎 ☠ 85cm

Trimeresurus wagleri, 🐀 🐦 🌀 ☠ 80cm

Trimeresurus wagleri, 🐀 🐦 🥚 ☠ 80cm

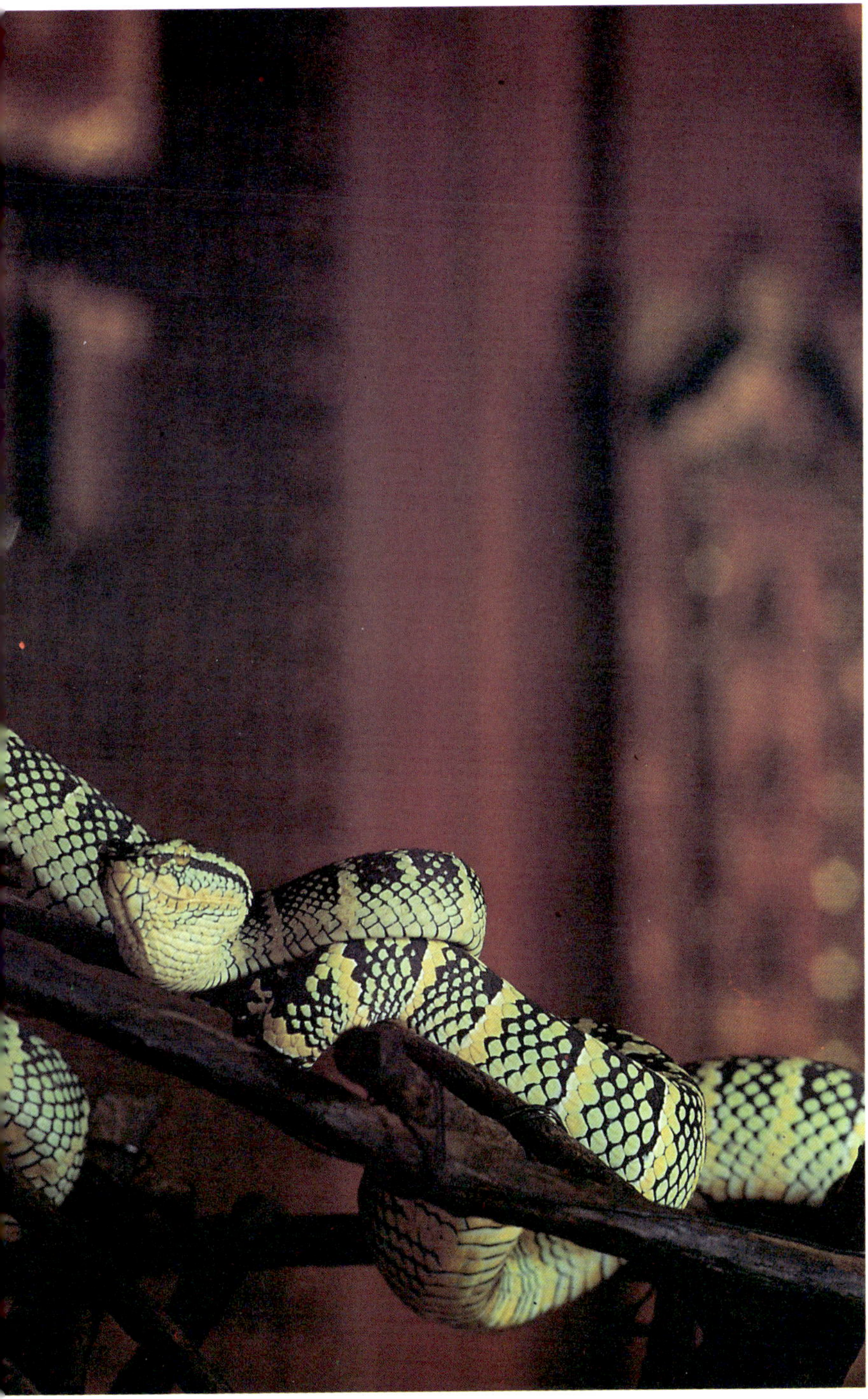

Trimeresurus popeorum, 🐀 🐦 🌀 ☠ 90cm

Trimeresurus popeorum, 🐀 🐦 🦎 ☠ 90cm

Trimeresurus wagleri, 🐀 🐦 🥚 ☠ 80cm

Snake Natural History

On the evolutionary scale, the snakes are the most recently developed herptiles. Since they evolved from lizards, some species actually still have vestigial limbs.

"The snakes, the last of reptiles to evolve, are in essence highly modified lizards."
Edwin H. Colbert, 1955

The natural history of snakes comprises factors of evolution, classification, and general biology, a basic knowledge of which will enable the snake keeper to develop a deeper understanding of the subjects of his chosen hobby.

Snakes form the suborder Serpentes within the order Squamata, which they share with two other suborders, the Lacertilia (lizards) and the Amphisbaenia (amphisbaenians). The Squamata is by far the greatest reptilian order and contains approximately 5800 living species, including about 3000 kinds of lizards, some 130 sorts of amphisbaenians, and about 2700 species of snake. It is almost impossible to give exact numbers of species in such great orders due to the frequent reclassification by taxonomists as further research is conducted and as new species are discovered or others become extinct. As man has reached to almost every corner of the globe, there are probably not many species of snake remaining to be discovered; but we still get the occasional pleasant surprises. By comparison to the Squamata, the other orders of living reptilians are relatively sparse in species: Testudines (turtles), with about 220; Crocodylia

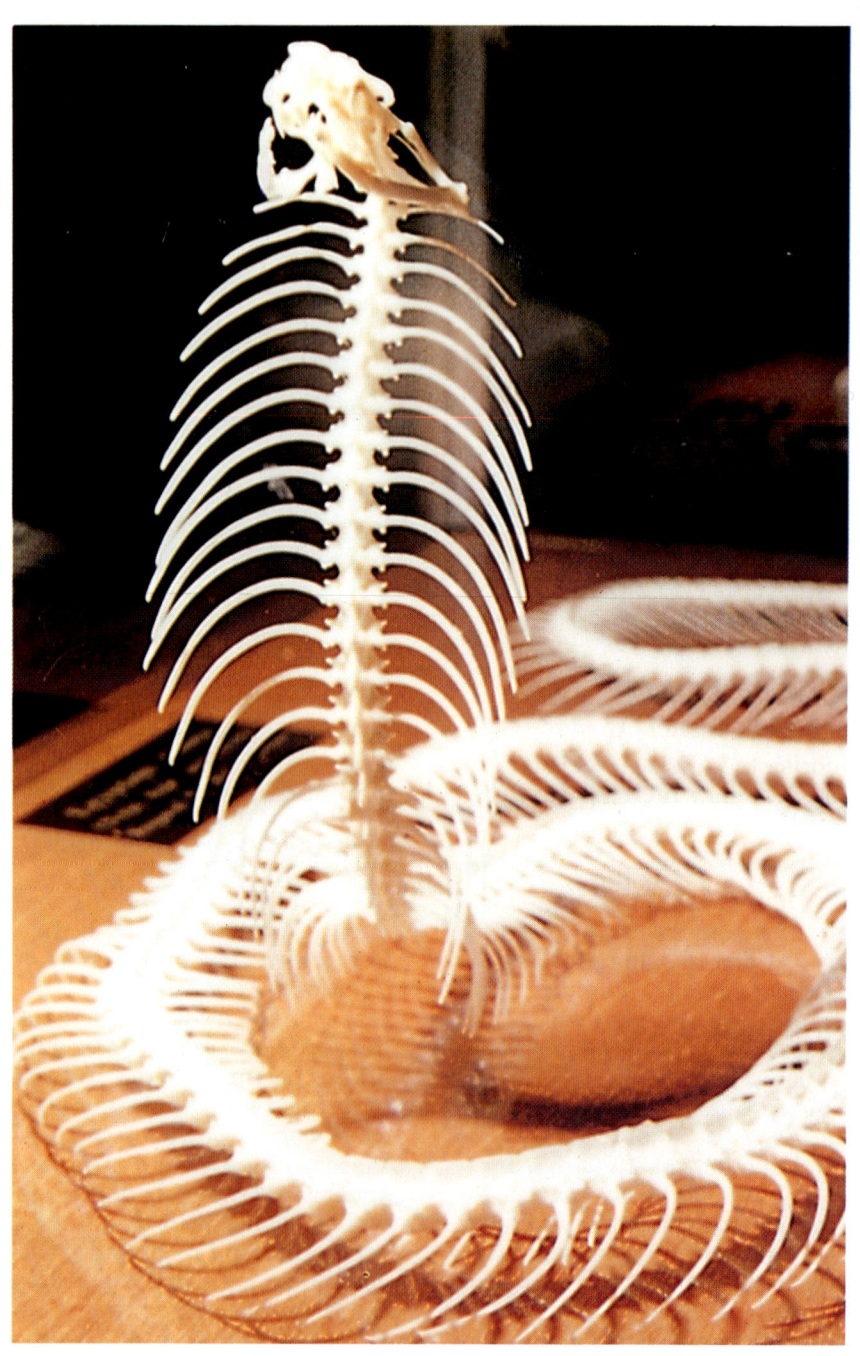

Snakes have been on the earth for over 120 million years. The elapids, a family of venomous snakes to which this cobra belongs, first appeared during the Miocene Period, about 12 to 30 million years ago.

(crocodiles, alligators, etc.), with 21; and Rhynchocephalia (the tuatara) with a single species. In evolutionary terms, the snakes form the most advanced group of the Reptilia.

Evolution

"The origin of snakes is a fascinating problem, though it is still largely a matter for speculation... ."
Angus d'A. Bellairs, 1957

To understand the position of snakes in zoological classification, their evolution must be discussed in relation to both the other reptilian orders and other classes of vertebrates. It is widely accepted that life began in the water and that it consisted of single-celled organisms which gradually evolved into more complex animal and plant forms. The process of evolution has been laboriously pieced together by scientists studying fossil remains of countless generations of extinct organisms. The first positive vertebrates (animals with backbones) were fishes, and for millions of years numerous fish species were virtually the rulers of the seas. Plants and some invertebrates were the first living things to venture onto the land, but it was not until the appearance of the crossopterygians, lobe-finned fishes, during the Devonian period (about 380 million years ago) that an opportunity arose for vertebrates to invade the land.

Eusthenopteron was a particularly important genus of lobe-finned fishes as it was on a direct line to the early amphibians. These particular fish were among the first to develop limb-like fins that allowed them to creep over the land to other sources of water. Additionally we can assume that the fish developed primitive lungs which enabled them to spend much longer periods in the atmosphere. The first actual amphibians evolved from these piscine ancestors about 35 million years later. Amphibians are today represented by frogs, toads, salamanders, newts and caecilians. One or more groups of these early amphibians were the ancestors of the reptiles.

Ideal conditions during the Carboniferous period which followed the Devonian were very favorable to amphibian proliferation. Extensive vegetated swamps were commonplace over much of the earth and the atmosphere was almost permanently moist, providing a perfect environment. The humid conditions allowed the amphibia to spend more and more time completely out of the water, where they could feed on the thousands of species of terrestrial invertebrate prey which hitherto had no vertebrate predators to contend with. Eventually, many amphibian species only returned to the water in order to spawn and lay their soft-shelled eggs, which hatched into gill-bearing, fish-like larvae. These developed into terrestrial forms by a process known as "metamorphosis," after a variable amount of time as a larva. This process is still retained by modern amphibians to this day.

The conquest of the land by vertebrates can be considered as one of life's greatest achievements, though the actual evolutionary processes involved were very gradual, taking place over millions of years. Every niche on the earth's surface capable of supporting life was eventually colonized and, as one niche was destroyed, another was left to be colonized by something else. However, the habitats of the amphibians were somewhat limited as they had to remain near water or at least in humid areas. In arid conditions, fluid loss through the naked, porous skin was rapid and often fatal.

To further exploit the terrestrial environment, more important evolutionary adaptations were required, the most essential being means of combating water loss. This was resolved by developing a protective cover to the skin, which eventually evolved into the typical reptilian scales. The development of the first hard-shelled eggs was taking place at the same time, and this meant that internal fertilization became necessary. The particular animals with these abilities were no longer amphibians, but had become primitive, true reptiles which could live and reproduce on the land. A major advantage was that eggs could be more safely concealed from predators on the dry land by burying them in the ground or hiding them away amongst dense vegetation. These large-yolked eggs contained enough nutrients to enable the embryos to develop to a relatively advanced stage before they hatched, giving them a greater chance of survival. The embryos still had to go through the early stages of development, of course, but they were now enclosed in a sort of sac, called the amniotic membrane. The embryo grew in the watery amniotic-fluid contained in the inner membrane, receiving oxygen from the outside and disposing of carbon dioxide by means of another membrane, the allantois. A further membrane, the chorion, surrounded the whole lot just inside the outer tough, protective shell.

The reptiles thus had a tremendous advantage over the amphibians, which had to spend all their time near water and lay their eggs in it. The reptiles continued to develop in various forms, each development being a further improvement for a terrestrial existence. Some

Snakes have evolved into having two methods of giving birth, neither of which is truly "superior" to the other. Above, the hatching of a Corn Snake, *Elaphe guttata*. Below, the live birth of a Copperhead, *Agkistrodon contortrix*.

improved their dry land hunting prowess, developing strong, muscular limbs which helped them pursue their prey speedily as well as escape from larger predators which were also developing. Others returned to the water to catch fish or amphibians in their strong jaws.

During the late Carboniferous period, the earth's climate began to change dramatically. Most of the typical swamps of the era dried up and many amphibian species became extinct. This was an advantageous period for the animals that were already terrestrial and laid shelled eggs. However, herbivores had yet to develop, and the reptiles had to feed on each other. Various adaptations for gripping, tearing and chewing had to evolve in this tremendously competitive time as thousands of different species developed. Legs became longer and moved below the body rather than at the sides, enabling animals to run faster and longer. Herbivores developed, and the first true mammals also appeared at this time (mammals are furred and suckle their young).

During the following Triassic and Jurassic periods the dinosaurs arose. These giant reptiles were the dominant creatures on earth for the next 140 million years before they eventually died out toward the end of the Cretaceous period.

Though fossil records of the Squamata are sparse, there is sufficient evidence to assume that the ancestors of our modern snakes and lizards were also making their appearance during these periods. The Lacertilia probably branched off from the primitive order Eosuchia during the Triassic period (240-200 million years ago), though the oldest direct ancestors of our modern lizards are known from fossils of the Upper Jurassic period (about 140 million years ago). Small lizards, and especially snakes, pose immense problems to paleontologists and, unlike the generally massive skeletons of the dinosaurs, are known only from scattered fragments of skulls and vertebrae. Piecing such fragments together is infinitely more difficult than the hardest of jigsaw puzzles, especially when most of the key pieces are missing. Much of our knowledge of the evolution of the Squamata remains, for the time being, necessarily sparse. But as research with modern equipment progresses, we will doubtless soon be able to find more positive solutions to those problems which are only partially resolved.

It is generally accepted that snakes evolved from lizards, but any direct connection is yet to be discovered. The term "missing link" is especially significant when applied to snake phylogeny, as only small numbers of fossilized snake genera have been found, and no intermediate species have been discovered. One reasonable explanation is that snakes evolved from burrowing lizards, which first lost their limbs, external ears, and, almost, their eyesight; all of which were of limited use below ground. During the burrowing period, the unique sense of prey detection, using the forked tongue and the organs of Jacobson, was improved dramatically, while possibly the beginnings of heat reception pits in some species began to develop. Certain species eventually ventured onto the land surface again. The lidless eyes redeveloped, but the limbs (though some of the more primitive snake families still possess vestigial pelvic girdles) and external ears were lost forever, being replaced by other sophisticated means of locomotion and hearing. Of course, it would have taken countless generations of reptiles over millions of years for these evolutionary processes to arrive at their current state. Today we still have burrowing snakes, limbless amphisbaenians which possess characteristics of both lizards and snakes, and we have the legless lizards as well. What we do not have is a concrete link between the suborders. So, for the time being, our knowledge of the path of snake evolution still remains uncertain.

Some snakes have developed odd and fascinating habits over the course of time. The hognoses, genus *Heterodon*, for example, have an interesting defense mechanism—they play dead.

Classification

When herpetoculturists start crossing snakes of one genera with those of another, the chances of classifying them properly become almost nil.

"The chief object of every branch of Zoology I consider to be the discerning of the species...."

Albert Guenther, c. 1858

The first generally accepted logical system of animal and plant classification had its origins in the first half of the 18th century. As European explorers reached the more remote parts of the continents and seas, scientists became increasingly excited (and confused) by the apparently infinite number of new animal and plant species being discovered. Zoological and botanical specimens were collected from all corners of the globe and sent back to European universities and museums where scientists had the task of examining them and trying to fit them into a category. It soon became apparent that a system of classification was required which was not only logical but international. In order to avoid costly duplication of work, it was important that the work of scientists in North America, Great Britain, or Germany, for example, could be communicated to those in France, Italy, or Sweden. Scientific literature, therefore, was usually presented in classical Latin as this had been the universal language of learned scholars for generations.

The Swedish biologist Karl von Linne (1707-1779) Latinized his own name to Linnaeus (by which he is usually described), then revolutionized our systems of classification when he perfected a binomial system of scientific nomenclature and published it in his *Systema Naturae*. The science of the theory, procedure, and rules of classification of organisms according to the similarities and differences between them is known as "taxonomy," and Linnaeus can be regarded as the father of this science. The binomial system consisted of applying a double Latin name to every species. The original system was of course primitive by today's standards, but it set a precedent for the following generations of taxonomists.

Natural classification is a hierarchal arrangement of animals or plants into different groups, based on differences and similarities among them. The lowest rank in this arrangement is the **species**, which is one of a group of organisms which are all essentially the same, at least with very little variation, and which freely reproduce with one another and create more individuals of a similar type. A number of species which are not essentially similar, but have several characteristics in common, are grouped into a **genus** (plural: genera). Numbers of genera are placed into **families**, families into **orders**, orders into **classes**, classes into **phyla** (plural of phylum), and so on. The number of similarities between members of a group becomes increasingly less at each step up the hierarchal ladder. For example, members of the same family have less in common than members of the same genus (unless, of course, they are in the only genus within the family). When difficulty arises in placing a species, additional categories such as subfamily or infraorder may have to be used.

Taxonomy is a complex subject that has caused much argument and controversy among scientists for generations. Modern classification is based on phylogeny (evolutionary history of species) as far as this is known, but in those groups where evolutionary information is sparse it is supplemented with a degree of calculated guesswork.

The binomial system has been much improved over the years. Each new species discovered is still given a two-part "Latinized" name, known as a "binomial." This is made up of the generic name and a specific or trivial name assigned by the taxonomist who gives it its first scientific description (not necessarily

Some taxonomists love to break down certain groupings into even further subgroupings. This *Amphiesma stolata*, for example, belongs to something called a subfamily (Natricinae), which is obviously a branch of the larger family classification.

the discoverer). Thus, in the species *Elaphe dione* (Steppes Ratsnake), *Elaphe* is the generic name which is applied to all snakes in that genus (all 35 or so of them) and *dione* is the specific name which is applied to a single species only. The name of the original author(s) and the year of description are written after the binomial. For example, *Elaphe taeniurus* Cope, 1860; *Crotalus adamanteus*, Beauvois, 1799; *Hapsidophrys lineata* Fischer, 1856. In cases where the species has been moved from its original genus, the specific author(s) of the original name are still retained, but in parentheses, for example: *Elaphe vulpina* (Baird & Girard, 1853).

Some species show slight geographical variation in scale counts, color, or form. When these variations are insufficiently different to warrant separate specific classification, they may be classified as subspecies. In such cases a third name is added to the binomial, making it a trinomial. Subspecies will interbreed quite readily and indeed do so at the borders of their individual ranges, producing examples with characteristics from both groups; such individuals are known as "intergrades." A good example of subspecific nomenclature is applied to the North American Common Kingsnake, *Lampropeltis getula*, of which many authorities recognize up to seven subspecies including the Eastern Kingsnake, *L. g. getula*; the California Kingsnake, *L. g. californiae*; the Florida Kingsnake, *L. g. floridana* ; and the Speckled Kingsnake, *L. g. holbrooki*. The first subspecies described has its specific name repeated in the trinomial (as in *L. g. getula* above), while further subspecies receive a new subspecific name placed after the specific name.

In scientific texts, generic, specific, and subspecific names are almost always written in a script different from that of the main text, usually italic. It will also be noted that scientific names are often abbreviated. When a species is first mentioned in a text, the full scientific name (with or without its author) is usually given, but when the same name is repeated it is abbreviated. For example: *Coronella girondica* is abbreviated to *C. girondica*, while *Drymarchon corais couperi* is abbreviated to *D. c. couperi*.

The following table demonstrates the classification of three snake species within the suborder Serpentes:

TABLE 1
Sample Classification of Three Snake Species

KINGDOM	Animalia	All animals	
PHYLUM	Chordata	Animals with notochord	
SUBPHYLUM	Vertebrata (Craniata)	Vertebrates	
CLASS	Reptilia	All reptiles	
ORDER	Squamata	Lizards, amphisbaenians, and snakes	
SUBORDER	Serpentes	All snakes	
INFRAORDER	Scolecophidia	Henophidia	Caenophidia
FAMILY	Typhlopidae	Boidae	Colubridae
SUBFAMILY	—	Pythoninae	Colubrinae
GENUS	*Typhlops*	*Python*	*Elaphe*
SPECIES	*T. diardi*	*P. regius*	*E. guttata*

Biology of Reptiles

The snake skeleton is, as you can see here, made up mostly of ribs and vertebrae.

Before studying the particular biology of snakes, it is desirable to have a little knowledge about the general biology of reptiles. All reptiles have certain biological similarities, and a knowledge of these will enable snake keepers to deal more adequately with the animals in their care. We know that the reptiles were the first vertebrates to become fully terrestrial. This was accomplished though a combination of internal fertilization through copulation and the development of the first shelled egg. These together constitute one of the great evolutionary advances. Without it, the birds and the mammals, including ourselves, would never have appeared on the surface of the earth. The egg alone, of course, did not prepare the reptiles for terrestrial habitation. Indeed, the egg was one of the later adaptations to occur. The whole animal gradually modified itself to suit its terrestrial existence, and many of the characteristics evolved have been passed on to the modern reptiles.

Reptiles are tetrapods (meaning four-limbed—

somewhat paradoxical when referring to snakes, but it is assumed that snakes evolved from four-limbed ancestors). The skin is composed of horny scales and is fairly waterproof, an important factor for water conservation. The head is typically supported by a relatively long neck. By articulation of the atlas vertebra with the condyle of the skull, the head can be moved from side to side or up and down. The brain is relatively small and encased in a solid bony skull.

There are three to six separate bones in the lower jaw. The teeth are located on the edges of the jaw bones, and sometimes (especially in some snakes) on the palatal bones. In most species the teeth are replaced several times in succession (polyphyodont). All reptiles possess a tongue that is typically (though not always) highly mobile and considerably extensible (particularly in snakes).

In many species the limbs are laterally oriented and typically end in five clawed digits, though adaptation has caused them to be partially reduced or absent in some lizards, all snakes, and most amphisbaenians (some primitive species, however, still possess a vestigial pelvic girdle).

Reptiles have functional lungs and have a three-chambered heart (four-chambered in the crocodilians). Like the other lower vertebrates, reptiles are poikilothermic and do not automatically retain a constant normal body temperature (as do the homoiothermic birds and mammals) but rely on the environment to thermoregulate.

Snake Biology

Snakes and lizards have colonized every suitable habitat with the exception of the polar regions. Various species have adapted to arid deserts and tropical rainforests; to temperate heathland, highland meadows, and evergreen forests; to fresh water or the oceans. Within these various regions, they have further manifested themselves as arboreal, terrestrial, burrowing, or aquatic forms. It is only the fact that reptiles are poikilothermic that has prevented them from colonizing regions where the permafrost precludes hibernation below ground.

Snake or Lizard

Though distinction of snake or lizard may at first seem very obvious, there are borderline cases which are difficult. Many lizards have evolved on a line parallel to that of the snakes, and in the process have acquired certain snake-like characteristics (loss of limbs for example), while certain primitive snakes have retained certain obvious lizard characteristics

Are these snakes? No, they're lizards; legless lizards to be exact. Above is *Ophisaurus compressus*, and below, the head of *Ophisaurus apodus*. These animals are living evidence that leglessness is not a reptilian characteristic possessed only by snakes.

(vestiges of hind limbs for example). The two suborders have indeed so many similarities that scientists have classified them together in the same order.

You can usually distinguish a legless lizard from a snake by examining the following points: first, ascertain whether movable eyelids are present; if so the animal is definitely a lizard—most legless lizards possess moveable eyelids but no snakes do. Next, look for external ear openings; if present, the animal is definitely a lizard—no snakes have external ear openings. Further points to examine include the tongue, which is invariably notched in legless lizards, rather than forked as in snakes, and the tail, which is often easily lost through autotomy whereas (with very few exceptions) snakes cannot voluntarily shed their tails.

Snake Anatomy, Physiology, and Behavior

The best known characteristic of the snake is that it has no legs. Another typical serpentine feature is the long flexible body with no obvious definition between head, neck, thorax, abdomen, and tail. The head and neck region is usually more or less distinct from the thoracic region in terms of diameter (the degree of distinction of the head from the body by the neck is a major guide to the identification of many genera or species). The main body is usually more or less circular in cross section, though it may vary to ovoid (vertically or horizontally) or triangular. The underside is usually flattened, and there is often a slight ridge along the flanks between the dorsal and ventral surfaces. The base of the tail is just posterior to the vent (the external opening of the cloaca), the common orifice for excretory and reproductive functions. The length of the tail is variable between species and, in some cases, between sexes.

The Skin: To protect against physical hazards such as mechanical injury, desiccation, and wear and tear, a snake's skin must be tough, particularly as a snake has a relatively large area of skin in contact with the surfaces over which it moves. The skin is formed from three layers, the innermost one being the thickest and containing the pigment cells. The thin middle layer consists of cells which are continually dividing and growing in a plane parallel with the surface of the body. These new cells eventually die and form the horny epidermis which totally encloses the body and typically consists of overlapping scales. The scales of a snake form an integral part of the skin, unlike those of a fish which are attached to the skin surface. If a cast snakeskin is examined, it will be seen that the thickened scales are joined together by a much thinner membrane. These membranes are folds in the skin between the scales, while the scales themselves are the parts which are presented to the exterior.

The scales on the dorsal surface are relatively small, while in most species the ventral scales are large, broad, and arranged in a single row from the throat to the vent. Those on the underside of the tail (subcaudals) may be arranged in a single or double row, depending on the species. The type, number and arrangement of scales on each particular species tends to be fairly constant and these points are frequently used as an aid to the identification and description of species. A subcaudal scale count may sometimes be used to identify the sexes, as male snakes often have longer tails and a greater number of subcaudals than females do.

Dorsal scales may be smooth, keeled, or granular, depending on the species. Smooth scales are usually shiny and iridescent and often appear to be wet, a fact that probably led to the myth that snakes are slimy. Keeled scales (with a raised central ridge) often appear to have a dry, matte finish. Granular scales, which only appear in the Acrochordidae, are very small, usually roundish, and do not overlap to the extent of other types of scales.

Many species exhibit minute, circular concavities, just visible to the naked eye, near to the tips of all the dorsal scales. Known as apical pits, these organs may be arranged singly, in pairs (most commonly) or rarely in groups of three or more. These cavities are formed by localized thinning of the horny scale-covering and the epidermal cells beneath are supplied with bundles of nerve fibers. Though the full function of these apical pits is still not thoroughly understood, it is suspected to be concerned with the sense of touch or temperature detection. Additionally, specialized tactile papillae may be found in clusters on certain scales on the head, chin, throat, or ventral region of many species. These are probably stimulatory organs used by courting snakes when the male rubs his chin over the female's body and attempts to get his vent in apposition to hers.

Shedding

The outer dead layer of skin, the **epidermis**, is periodically shed, sloughed, or cast off as it becomes worn and as the snake grows. The cells of the middle skin layer form the replacement epidermis. When this new skin is fully formed, a lubricant fluid is

The shedding of a snake's skin is known as *ecdysis*. Above, a hognose snake, *Heterodon* sp., displays the opacity of the eyes. Below, the process of ecdysis is being aided by a snake's keeper with the help of a pair of tweezers.

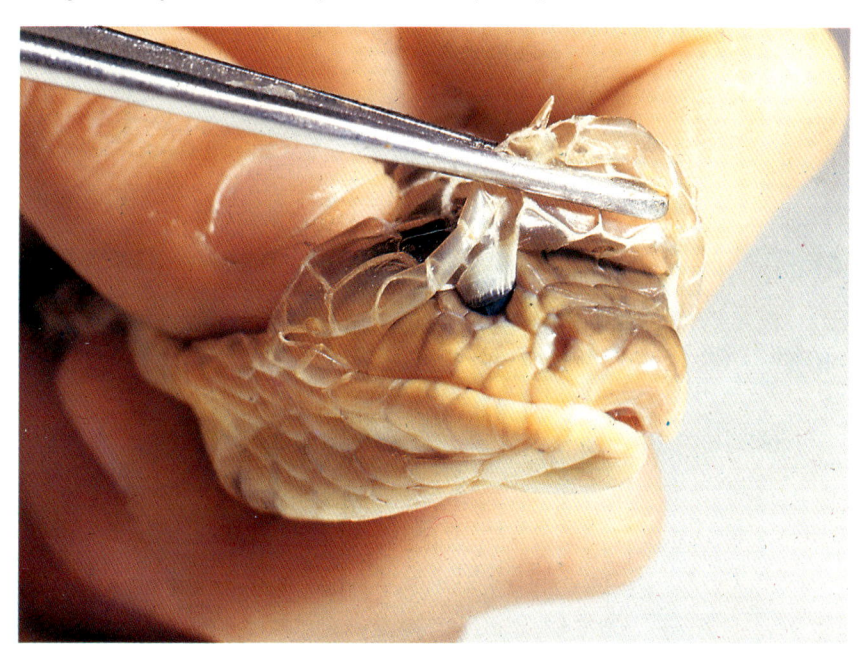

produced to separate it from the old skin. This causes the old skin to acquire a dull appearance and the sleekness associated with a normal, healthy snake is temporarily lost. The eye scale or **spectacle** becomes affected, takes on a milky appearance, and the snake becomes partially blind. During this time snakes seek seclusion, refuse to feed, and are often more irritable than usual, so precautions should be taken (especially with venomous species) to avoid the increased likelihood of bites. The eye usually clears again just before the shedding. As a rule the skin is first loosened at the lips and the snake actively seeks out rough surfaces on which to rub its snout. The loosened skin is turned inside out and pushed back over the head and body, somewhat in the manner of a lady peeling off her nylons. To facilitate the loosening of the skin, the snake will expand and contract its muscles in shivering waves and perhaps attempt to crawl over or between rough-surfaced objects so that purchase is applied. The skin should shed in a single piece if there are no health or environmental problems. The cast skin is almost transparent and shows only a trace of the pigmentation. However, the scales can still easily be seen and shed skins are often used to identify the presence of species in certain habitats.

Juvenile snakes usually shed their skins two to ten days after birth or hatching and almost invariably before their first meal. A snake may shed up to ten times in its first year, when growth is rapid. This number reduces each year as the rate of growth decreases. Adult snakes usually shed no more than three or four times per year. An unhealthy snake may have difficulty in sloughing its skin completely, which can cause further health problems.

Locomotion

One cannot help but marvel at how the limbless snake can move with such grace and agility. The flowing locomotion of most snakes is made possible by the special vertebral joints that ensure the reptiles can bend considerably in almost any direction. In most species, locomotion is achieved by lateral undulations of the dorsal muscles coupled with rectilineation, though each method can be used individually. The lateral undulations of the body allow the scales along the flanks to come into contact with imperfections or objects on the terrain, which is why most snakes prefer to move through undergrowth or rocky terrain rather than over a smooth surface.

When moving slowly or pensively (as when stalking

Arboreal (tree-dwelling) snakes have modified muscular abilities that enable them to negotiate trees, shrubs, and so forth with relative ease. In snakes that are only partially arboreal, like this Boa Constrictor, *Boa constrictor*, for example, such modifications are not as well-developed.

prey), however, rectilinear locomotion alone may be used. In such a case, the snake moves in a more or less straight line using the broad ventral scales (or gastrosteges) that are attached to the snake's numerous rib pairs by special muscles. The ribs articulate with the vertebrae and, if one could imagine a snake's skeleton moving, the ribs could be compared with the legs of a centipede. Simultaneously, the muscles expand and contract and cause the broad ventral scales to move in waves, the posterior edge of each scale pushing against imperfections in the substrate and causing the snake to go forward. Thus if a snake is placed on an unnaturally smooth surface (a sheet of glass or a polished linoleum floor for example) it will have obvious difficulty in moving forward. Weighty snakes such as pythons, boas, and large viperids use rectilinear crawling almost exclusively—as if their obesity has caused them to become lazy. Only if alarmed or if swimming is a weighty snake likely to break into undulatory movement. Some snakes can also extend and contract their bodies a little in a sort of concertina movement which may be used in conjunction with rectilineation.

Arboreal snakes, or those which frequently climb, have further adaptations to help them move through vegetation. The lateral scales are often heavily keeled and the ventrals sharply angled at the edges to help them get a grip on bark, etc. Some snakes have an amazing capability of being able to scale a fairly smooth tree trunk vertically by using these adaptations. The flying snakes, members of the genus *Chrysopelea*, which would probably be better called gliding snakes, have taken their locomotory proficiency a step further. When pursuing prey, or avoiding predators, they may launch themselves into the air and paraglide for some considerable distance before landing at a lower level.

Aquatic snakes use undulatory movement to swim. Some of the sea snakes even have laterally flattened bodies and paddle-like tails to increase their swimming efficiency. Most true aquatic snakes have more or less replaced the broad ventral scales with smaller scales resembling those on the dorsal surface. Most species of snake appear to be able to swim quite efficiently should the occasion arise.

Some desert-dwelling species that inhabit terrain with a loose, shifting substrate have developed a specialized means of locomotion known as sidewinding. Being unable to efficiently use more coventional means of snake locomotion, they have developed a system of throwing body loops, so that the main pressure of movement is directed vertically downward. The snake moves at an angle of about 45 degrees to the line of its own body by raising the head and throwing it forward in the direction of travel. As the head contacts the ground, a loop of the body is raised and moved toward the head, which by this time is beginning its next step. Snakes moving in this manner leave characteristic J-shaped tracks in the soft substrate.

Snakes from various parts of the world, but in similar habitats, have developed this method of locomotion. The Sidewinder, *Crotalus cerastes*, of S. W. USA and Mexico, the Saw-scaled Viper, *Echis carinatus*, from N. and E. Africa and S. W. Asia, and the Dwarf Puff Adder, *Bitis peringueyi*, from Angola and Namibia, are all adept sidewinders and all inhabit areas where this system of locomotion is necessary. Other snakes occasionally sidewind to a certain extent if the occasion arises and sidewinders themselves are also able to move in a more conventional manner.

Sea snakes are probably the least known and studied of all the serpents. Their venom is highly potent and they occur in areas where most scientists cannot get to them easily.

Speed

Much has been conjectured on the apparent speed of snakes. A snake being able to overtake a racehorse is one of the myths frequently heard on the subject. Disappointingly, perhaps, no snake would be capable of such a speed. Indeed, no snake would be likely to outdistance a man walking briskly. As an example, perhaps the fastest North American snake, the harmless Coachwhip, *Masticophis flagellum*, has been clocked at a speed of 3.6 m.p.h. and it is estimated that its maximum capable speed would not exceed 7 m.p.h. Even if it could reach such a speed, the metabolic rate would ensure that no snake has the ability to maintain it for more than a few seconds.

The Senses

Many people are surprised to hear that snakes are extremely sensitive animals that possess all of the senses with which we ourselves are blessed, though some are less efficient, and others more efficient, than ours. Let us examine the basic senses of snakes and compare them with our own.

Sight: A typical feature of snakes is that their eyes have no moveable eyelids, but the cornea is protected by a circular, transparent scale (or, in some cases, a transparent part of a larger scale) known as the brille or spectacle. Considering their predatory nature, the eyes of snakes are relatively inefficient, but like other animals with poor eyesight, they make up for this apparent disability by having other extremely sharp senses.

A snake's eyesight can be considered inferior when compared with ours or even with that of most lizards. It seems to have an inability to focus, and its perception of detail is apparently poor. With a few exceptions (*Ahaetulla* and *Thelotornis* for example) snakes have difficulty in recognizing stationary objects by sight and must rely on the perception of movement of living prey or use other senses to detect dead or immobile prey. Eye size and efficiency depend on the species' methods of catching prey. Most burrowing species have eyes in various stages of degeneration, while many arboreal and fast moving, diurnal terrestrial species which actively hunt by sight have large eyes; most of these also have circular pupils. A few of the nocturnal, arboreal genera (e.g. *Hapsidophrys*) have extremely large eyes with round pupils, but the majority of the nocturnal genera exhibit vertically slit or elliptical pupils. There are exceptions in that some species with vertical pupils are diurnal, but it is probable that these have evolved from nocturnal ancestors.

Hearing: Snakes do not possess external ears but they apparently can hear as they react to noises of varying degree and pitch. However, the snake's auditory sense is probably far from efficient when compared with ours or even with that of most lizards, but it is efficient enough for snakes to be successful. In normal hearing, airborne sound waves cause the typanum (ear drum) to vibrate and the sound is

The common name for many species of *Coluber* is "racer," and refers to these snakes' relatively remarkable speed. Ironically, it is generally believed that no snake can attain a speed much over 5 mph for more than a few moments.

Members of the family Crotalidae are known as the "pit vipers," because of the infrared receptor pits located on each side of their heads between the eye and nostril. On this Pigmy Rattlesnake, *Sistrurus miliarius*, you can see this pit quite clearly.

thus transmitted to the brain via the ear bones and nerves. A snake hears by picking up vibrations from the solid surfaces on which it rests. The vibrations pass through the snake's jawbone to the quadrate and then to the stapes which is in close proximity to it. From there, nerve impulses are transmitted to the brain in the usual manner. The lung may also be involved in the transmission of sound and other vibrations.

Smell: Although snakes have a normal sense of smell operated by the sensory epithelium lining the nostrils, this is relatively poorly developed. Snakes mainly rely on what, to us, can be regarded as a roundabout way of detecting odors. Situated in the palate, just below the nose, and corresponding to but not adjoining the nostrils, is a pair of domed cavities lined with sensory epithelium which is connected to a special branch of the olfactory nerve. These vomeronasal pits are known as Jacobson's organs and are used to detect scent particles which are introduced to them by the tongue. It is believed that the tips of the snake's highly mobile forked tongue are designed to fit into the organs and would explain the continual flickering of an active snake's tongue, which it protrudes through a labial notch (making it unnecessary to keep opening the mouth). The scent particles are picked up from the air or solid objects on the tips of the tongue and transferred directly to the inner surfaces of the Jacobson's organs from whence the necessary messages are passed to the brain. This sophisticated sense of smell is highly important to snakes, who use it to find prey, seek out a mate, or to detect potential enemies.

Balance and Touch: A snake has an efficient sense of balance which is controlled by the semicircular canals of the inner ear. These respond to head movements as well as the muscles affected by changing gravitational stresses as other parts of the body are tilted. Thus, in spite of its long cylindrical shape, a snake is able to stay the right way up no matter what situation it finds itself in.

A snake also possesses a highly efficient sense of touch. A light touch to any part of a resting captive snake will produce a reaction ranging from a twitch of the area touched to an angry defensive action or attempt at flight, depending on the species of snake and its degree of tameness. This indicates that the snake's skin is amply supplied with sensitive nerves. Snakes are thigmotactic in that they respond to the stimulus of closeness to solid objects and feel uneasy when caught out in the open with little opportunity of taking refuge. When at rest most snake species squeeze themselves tightly into whatever cavity is available. This is an important factor in captive snake husbandry. Many snakes which are not provided with a refuge into which they can tightly squeeze will suffer stress, refuse to feed, and slowly deteriorate in health. The Royal Python, *P. regius*, is a typical example of a snake which will steadfastly refuse to feed unless it has the facility to secure itself in some tight cavity.

Heat-sensitive Organs: These highly specialized organs are found in only two snake families. In the Boidae, all members of the subfamily Pythoninae (except *Aspidites* and *Calabaria*) and the genera *Corallus*, *Epicrates*, and *Sanzinia* of the subfamiliy Boinae have these organs arranged in series along the jaw-line, either within the labial scales (Pythoninae) or between them (Boinae). All members of the family Crotalidae (also known as pit vipers) possess a single pair of prominent pits, each located between the eye and the nostril. These heat-receptor pits contain a membrane which is extremely sensitive to temperature changes and snakes possessing them are able to locate the direction of warm-blooded prey and strike at it accurately even in complete darkness.

Internal Systems

Compared to most other animals, snakes are an odd shape. But they still possess all of the internal organs that are necessary for basic physiological functions. These, of course, have become adapted in one way or another to fit into the elongate shape. Let us make a brief examination of the snake's internal physiological systems.

Feeding and Digestion: The mouth cavity contains various glands to help moisten and lubricate the prey in preparation for swallowing. As snakes have no means of tearing, crushing, or chewing. Their relatively large prey must be swallowed whole, so adequate lubrication is a very important factor. When the snake is stimulated by the prospect of a meal (by a combination of the olfactory, visionary, and/or heat receptory senses), the oral glands produce an additional quantity of lubricatory mucus (in other words, its mouth waters).

In poisonous snakes, the venom glands are modifications of a pair of salivary glands which produce normal saliva as well as venom from the Duvernoy's glands (sometimes called parotid glands), which are incorporated within and at the rear of the upper labial

One of the best-known abilities snakes have is that which allows them to swallow large food items whole. This "skill," you might say, is well demonstrated in this photo.

salivary glands. There is one situated on each side of the upper jaw, usually just behind the eye, often extending well into the cheek and, in a few genera, even into the neck and/or thorax. The venom gland opens into a duct which passes venom through the canal or along the groove contained in the venom fang.

The Teeth: Snake species show varying degrees of tooth specialization. Teeth may be carried on the the dentaries of the lower jaw, and the maxillae, the pterygoids, and the palatines of the lower jaw. There are exceptions. In primitive snakes of the family Leptotyphlopidae, for example, teeth are absent from upper jaw, while members of the family Typhlopidae have no teeth in the lower jaw.

Snake teeth are generally sharply pointed, strongly recurved, and fused to the bones on which they are affixed. At regular intervals, the older teeth are shed and replaced by new ones. The cycle of tooth replacement ensures that matured teeth and partly grown teeth of various sizes are always available and well-distributed on all the tooth-bearing bones.

Most of the more typical non-venomous snakes have teeth of a fairly uniform size, though those at the front or the rear part of the maxilla are sometimes larger.

Snakes bearing these simple teeth are known as **aglyphic** species. Some subfamilies of the Colubridae (Boiginae, Homalopsinae, Aparallactinae, Elachistodontinae, Xenodontinae, and Natricinae) contain species which are known as rear-fanged or **opisthoglyphic** species. In such cases, two or more fangs at the rear of the maxilla are enlarged and are grooved (usually anteriorly) so that venom can be conducted from the venom gland and introduced into the prey.

In species of the families Elapidae and Hydrophiidae the venom fang is canaliculate (really an adaptation of the groove, as the suture or join can be seen along the anterior surface of the fang) and situated at the front of the maxilla. The pair of fangs is fixed more or less rigidly on the maxillae. Such an arrangement is known as **proteroglyphic**. When the snake bites, the contents of the venom gland pass into the canal of the fang and are injected into the body of the prey via an aperture near to the point of the fang. In a few genera (*Hemachatus*, for example) the aperture is situated at the front of the relatively short fang, and by rapid contraction of the venom gland, venom can be sprayed directly at an aggressor, usually in the direction of its eyes. The aim is fairly accurate up to two or three meters (6-10 ft) and venom in the eyes can cause severe pain, inflammation, and temporary blindness. Goggles should be worn when dealing with such species.

Members of the families Viperidae and Crotalidae, known as the **solenoglyphic** species, have the most advanced kind of venom apparatus. The hypodermic-like fangs have become almost completely canaliculate and are the only teeth present on the maxillae. Unlike the relatively fixed fangs of the proteroglyphs, they lie back along the roof of the mouth during normal activity, but can be rapidly brought forward by rotation of the maxilla. The fangs of solenoglyphous snakes are usually relatively long, those of a large Gaboon Viper, Bitis gabonica, being up to 2.5 cm (1 in) in length. When striking, many solenoglyphs open the mouth so wide that the angle of the jaw almost reaches 180 degrees. The fangs are then directed at the prey in a stabbing rather than biting action, and the mandibles are closed only after the fangs have gained contact.

Dietary Range: Snakes are exclusively carnivorous reptiles that may take vegetable material only accidentally during prey consumption. Many juvenile snakes, as well as adults of the smaller species, eat invertebrates, including worms, molluscs, insects, and spiders. There are even small species which specialize in eating particular invertebrates, such as termites, crickets, centipedes, or snails. Larger snakes feed almost exclusively on various vertebrates; some may feed indiscriminately on whatever is available, though the vast majority of snakes tend to specialize in certain groups of animals that are partially determined by the range of prey available to them in their particular habitat. Thus snakes living in or near water feed predominantly on fish or amphibians; desert living species feed principally on lizards and small mammals; while arboreal snakes may specialize in feeding on geckos, tree frogs, or nestling birds. Two genera of snakes, *Dasypeltis* of Africa and *Elachistodon* of India, feed exclusively on the eggs of birds, while there are even those which feed exclusively on other snakes, including the King Cobra, *Ophiophagus hannah*.

Catching and Killing Prey: Snake species have various means of acquiring a square meal; there are those which actively go out hunting in locations where suitable prey is likely to be found, others which lie passively in ambush waiting hopefully for their prey to pass by, and those which use a combination of these two methods. A few species may use the tip of the tail as a lure to attract small predators, but then they themselves may become the prey because of this.

A common and simple means of catching prey is the "grab and swallow" technique. After locating the prey animal by sight, smell, or a combination of both, the snake grabs the prey in its mouth. Once the prey is trapped by the recurved teeth, the snake begins to swallow it. Many of the primitive burrowing snakes as well as many colubrids, particularly fish- and amphibian-eaters, feed in this way.

Constriction, as used by many colubrids and the boids, is another method. The prey is grabbed in the mouth and a reflex action immediately follows as a number of coils of the snake's body are rapidly wound around the prey, quickly immobilizing it. Sufficient pressure from the muscular coils is applied to suffocate the prey animal fairly rapidly but not, as is commonly believed, to crush every bone in its body. Once the prey is secured in the coils, it is held in such a way that the prey cannot use any defensive organs and damage the snake. Once the prey is dead the snake can leisurely swallow it, usually starting at the head end.

The most sophisticated means by which snakes overpower their prey is by envenomation. A venomous snake introduces sufficient poison into its prey to kill it or at least to paralyze it. In some species of rear-fanged snakes, elapids, and hydrophiids, the grab and swallow technique is used with the added advantage of the prey being immobilized by envenomation. Some elapids, most viperids, and crotalids envenomate the prey and quickly release it before it can retaliate. In such cases the prey, though already weakened, often covers a considerable distance before it becomes totally immobilized by the effect of the envenomation. After a few minutes, the snake will track down its victim using its olfactory senses before swallowing it in the usual manner.

Swallowing: Snakes are famous for their ability to swallow prey several times larger than their own heads. There are, however, many grossly exaggerated stories with regard to the size of prey swallowed by some of the larger boids. Most snakes have an extremely large gape and recurved teeth that allow them to seize and retain a grip on relatively large animals. The bones in a snake's jaws and palate can be moved independently and are connected by very elastic tissue which allows the buccal cavity to enlarge considerably during the swallowing process. The

The pride of many snake keepers is a specimen that will take food right from the hand.

lower jaw may also be thrust forward, and each side of the jaw is capable of independent movement. The two halves of the lower jaw are connected by an extremely elastic ligament at the chin, thus allowing the dentaries to spread a considerable distance apart. As the snake clamps its mouth shut over the prey, the movable bones reverse and begin to draw the prey into the buccal cavity. By alternating movements of the various bones on each side of the jaw and using the elasticity of the skin and muscles, the snake is gradually able to work prey into the gullet, where peristaltic action takes over and the food makes its way to the stomach.

In contrast to the highly mobile bones of the jaw and palate, those of the cranium form a very firm and rigid container for the brain, which protects this delicate organ from pressure and damage during the process of swallowing.

Snakes generally secure prey of a size range suitable for easy swallowing and there is a lower as well as an upper limit of size. While a captive Boa Constrictor, *Boa constrictor*, for example, will feed eagerly on mice as a juvenile, there comes a time when it will find such small prey difficult to deal with and graduate to rats, rabbits, or chickens. There is no reason to doubt that similar behavior is practiced in the wild.

Frequency of Feeding: As poikilothermic animals have a relatively slow metabolic rate and a correspondingly low energy requirement, most reptiles feed infrequently when compared to birds or mammals. Additionally, most snakes take large meals of whole prey animals that contain all of their nutritional requirements, and such a meal will last the snake some considerable time. It may come as a surprise to some to learn that many snakes feed every two or three weeks only, and at certain times of the year they may fast for even longer periods. Hibernating snakes will obviously fast for several months.

Young snakes needing to grow rapidly will usually feed more often than the adults, maybe two or three times per week for the first couple of months, then the frequency of feeding reduces as the reptile ages. Other factors which affect frequency of feeding include temperature, types of food (small, invertebrate-eating snakes feed more frequently than those feeding on vertebrates), and periodic sloughing.

Digestion: The snake has a digestive system similar to ours, but, of course, it is adapted to fit into the elongate shape. The relatively long esophagus has many longitudinal folds, allowing for a high degree of distension. Peristaltic action in the esophagus is poorly developed, and it is mainly the muscles of the neck and body which push the food toward the stomach. The stomach also has numerous longitudinal folds which give it an exceptional capacity for expansion. The main digestive process takes place in the stomach as the food is churned by muscular action.

A high proportion of hydrochloric acid in the gastric secretions helps digest the bones of the prey as well as the softer parts. The amount of time taken for complete digestion is variable and depends on the type of prey and the temperature. In optimum temperatures the prey may be digested in three or four days; in cooler conditions a week or more. A large amount of gas is often produced in the stomach during digestion and this may further increase its distension; the gas dissipates as digestion proceeds. When the food has decomposed into a semi-fluid mass, it passes into the small intestine, which is arranged in a great number of coils. The food is further broken down here and the nutrients are absorbed via the intestinal walls into the bloodstream. The undigested residue passes into the wider large intestine where more fluid is extracted and it solidifies. This fecal matter is compressed into solid pellets and stored in the cloaca at the base of the large intestine until defecation takes place.

The cloaca is the final part of the digestive tract. The interior of the cloaca is

When feeding dead food to venomous snakes, it is both logical and rational to use long forceps.

divided into three sections: the **coprodeum** where feces is gathered together; the **urodeum** where the urinary and sexual passages enter; and the **proctodeum** which leads to the exterior via the vent and lies transversely across the body. In female snakes the urinary and sexual passages are separate, but in male snakes they are joined to form a common channel emptying into the urodeum. Various anal glands which produce often foul-smelling secretions are situated at the junction of the urodeum and proctodeum.

Excretory System and Water Balance: A snake has to take in water to counteract fluid losses through the actions of respiration and excretion and to a lesser extent by evaporation through the skin. The amount of water required by snakes varies from species to species. Those existing in arid desert regions are more adept at conserving water and gain most of the fluid they require from that contained in the bodies of their prey. Snakes inhabiting less arid regions may drink regularly and frequently. They immerse the mouth in the water and draw the fluid in through the labial notch by expansions and contractions of cheek and neck muscles. Arboreal and desert species often drink dew and condensation which is deposited on foliage, rocks, or even on their own bodies.

Water balance regulation in snakes is performed chiefly by the kidneys, though the main function of these organs is to remove the nitrogenous products of metabolism. In reptiles (and birds), the liver converts the toxic ammonium compounds produced from the breakdown of proteins into relatively harmless uric acid. This uric acid, which is almost insoluble in water, is transported to the kidneys in the bloodstream. It is not excreted from the body in solution, as this would result in an enormous water loss. Much of the water is therefore recycled into the body by the kidneys and the residue is passed to the cloaca, where yet more water is removed. The final product, almost pure crystalline uric acid, is a fairly solid, whitish mass containing very little water. This is passed out through the vent, often together with dark-colored intestinal feces. This explains the dark- and light-colored parts in the droppings of reptiles and birds.

The Respiratory System: The left lung in most snakes is either vestigial or absent, while the functional right lung is elongate, usually extending about half way into the trunk. In some water snakes, however, it may reach back almost as far as the vent and acts additionally as a hydrostatic organ. Additional respiratory surfaces are provided by the vascular lining of the lung, extending forward onto the roof of the windpipe. In view of the snake's method of feeding, there is a modification of the glottis (entrance to the windpipe) which allows it to be protruded forward over the tongue during the swallowing of large prey. The glottis, like the rest of the trachea, is reinforced with rings of cartilage that prevent it from collapsing under pressure. The snake is thus able to take a protracted bulky meal without danger of suffocation.

Heat Balance and the Circulatory System: A snake's blood has generally much the same functions as that of other animals. It is the medium of gaseous interchange, provides a means of transporting nutrients to the tissues and removing waste materials, and plays an important part in thermoregulation by transferring warmth from the body surface to the various internal organs. As in all of the vertebrates, the snake's heart pumps the blood around the body through a system of arteries, capillaries, and veins. The snake's heart has two auricles and an incompletely divided ventricle so that in section it appears to have only three chambers (as

Many people wonder how snakes manage to swallow whole prey without cutting off their air circulation. This is done with the aid of a small modification of the glottis that reaches over the tongue. In essence, the windpipe is then "extended." The snake shown here is a Rough-scaled Bush Viper, *Atheris squamiger*.

Snakes are naturally attracted to the heat of the human body, which is why many of them will actually curl up in their keeper's hands and calmly remain there. The species the girl is holding is *Python regius*, otherwise known as the Ball Python.

opposed to four in the crocodilians and the higher vertebrates). Thus the pulmonary and systematic circulations are only partially separated.

As in all lower vertebrates, snakes are ectothermic (relying on external heat sources) and poikilothermic (having a variable body temperature). Unlike birds and mammals, snakes are unable to maintain a suitable body temperature through normal metabolism. At reduced temperatures, snakes become torpid and cannot respond fully to stimuli until they reach an adequate temperature. The average snake must maintain its body temperature between 4 and 38 degrees C, regardless of climatic conditions. The reptile may be able to survive temperatures just outside this range for short periods, but protracted periods of unsuitable temperatures will result in death or, at the very least, severe injury. Most snakes have a preferred optimum temperature for normal activity and in most species this ranges between 18-35 degrees C, although, of course, there are extremes, depending on the species and its habitat.

Being unable to control body temperature by physiological means, snakes must rely on behavioral methods, and the habits of various species differ considerably depending on the climatic conditions. In equatorial conditions for example, where temperatures remain fairly high and constant day and night throughout the year, there is no period of winter hibernation, and the chief hazard is overheating in the midday sun. In such cases, many species are nocturnal or burrowing, while those which are diurnal operate in the shade or during the hours of dawn and dusk (crepuscular). At excessively hot or dry times of the year, many species estivate and become inactive for varying periods.

At higher latitudes and altitudes, snakes face the problem of greater ranges of temperature between night and day, and more dramatically between summer and winter. In winter, the environmental temperature may remain below the snake's activity levels for several months, in which case the reptile can only survive in a state of torpor in a frost-free subterranean refuge.

Snakes must therefore seek out situations where they are able to operate at their preferred temperature. Snakes from colder climates will bask in the sun to reach their optimum activity temperature, though they will return to the shade to avoid overheating. Diurnal snakes may become partially nocturnal on warm summer nights, while some nocturnal species may warm themselves at night by lying on rocks or even roads which have retained heat from the daytime sun.

In conclusion, snake species are generally unable to satisfactorily adapt to conditions much different from the climates of which they are native, and this is a very important point to consider when we are setting up housing for captive specimens.

As a general rule snakes thrive in warm climates, but some species seem to have adapted to cooler regions as well. The garter snakes, *Thamnophis* spp., for example, reach all the way into central Canada.

The Terrarium

The setup shown here is both attractive and practical for many snake species. The snake moving about in the lower left corner of the cage is the popular Sinaloan Milk Snake, *Lampropeltis triangulum sinaloae*.

A snake terrarium once consisted of a simple, glass-fronted box heated with an electric light bulb. The light was often left on day and night, week in and week out, so that the reptiles were subjected to constant high temperature, bright light, and low humidity. Although many species may have seemed to do well in such terraria at first, the unnatural conditions would sooner or later contribute to the demise of the reptiles.

By applying our knowledge of natural ecology to terrarium science, successful snake keeping and breeding have come within the reach of anybody with an interest in this fascinating group of animals. It is now possible to purchase ready-made terraria with electronically controlled life support systems. However, many enthusiasts like to build terraria themselves, the advantages being that they can be made to any individual pattern and usually are much more economical. A great choice of construction materials is available and, with a little artistic imagination, a terrarium can be made into an extremely attractive feature or focal point in the modern home. The types of materials used will reflect the habits of the species to be kept; timber, for example, would be unsuitable for an aqua-terrarium or for snakes requiring an excessively humid environment, but is ideal for desert-dwelling species.

The shape of the terrarium is immaterial as long as it provides the right conditions for the species to be housed. Arboreal snakes, for example, would require a tall, roomy terrarium with facilities to climb, while a long, relatively shallow terrarium would be more suitable for terrestrial species. Active snakes such as racers require more space to move about than languid ones like pythons or puff adders. An aqua-terrararium is necessary for semi-aquatic species (such as many water snakes of the subfamily Natricinae), while totally aquatic genera, such as *Erpeton* and *Acrochordus*, require an aquarium containing water with little or no land.

Most snakes seek out confined spaces in which to rest and, when not actively hunting or breeding, will spend most of their time in a very small area. Not having to hunt over large areas for food, a captive snake requires relatively little living space.

Remembering that most reptiles find it difficult or impossible to adjust to climates strange to them, the wild habitat of the species being kept will dictate the conditions required in the terrarium. To decide what kind of accommodation is suitable for a certain species it will be necessary to obtain information on the habitat from a climatic atlas, from herpetological literature, or from a fellow herpetologist who is successful with the species in question.

Glass Terraria

Silicone rubber sealing compounds of the type used for sealing joints around glass windows, etc., are invaluable materials for the terrarium as well as the aquarium keeper. It is relatively inexpensive, and sheets of glass can be cemented together with it to produce a strong, watertight container of almost any desired shape or size. Moreover, sections of glass can be left out at the ends or the back so that ventilated access doors of drilled plywood, plastic sheeting, clear acrylic, or framed gauze can be fitted. The sealer is supplied in tubes which can be fitted into a caulking gun to make application easy. It may be obtained from aquarist's suppliers or hardware stores.

For small to medium snakes the terrarium is best constructed from 6 mm (quarter inch) thick tempered glass. Unless you are adept at glass cutting, it is best to get your glass dealer to cut the sheets to size for you. At the same time, you can ask him to grind smooth the edges which will be left exposed after construction (such as around the top) to eliminate the risk of cut fingers. When working out the dimensions of the glass sheets, be sure to remember the thickness of the sheets to allow for jointing. If using sealing compounds, be sure to follow the manufacturer's instructions, especially with regard to ventilation. The strong acetic acid vapors from the sealing compound can be unpleasant and overpowering.

The lid for a glass terrarium is best made from plywood and is constructed so that it just fits over the top of the main container. The depth of the lid will depend on the type and size of heating and lighting apparatus to be used, but 20 cm (8 in) should be adequate. Holes are drilled in the top of the lid for ventilation purposes. The lid can be painted, polished, or varnished to make it look more attractive.

Acrylic sheeting may also be used for terrarium construction, but a problem with this material is that it scratches easily and will not stand up to regular cleaning without an opaque film of fine scratches forming on the surfaces. However, an acrylic sheet in the back or end of a glass terrarium will allow you to include ventilation holes at locations suitable for efficient air circulation.

Today there are a number of companies producing modern-looking cages for pet snakes. Many of these cages are made of a durable plastic, which makes them easy to both lift and clean.

Fiberglass Terraria

Fiberglass terraria are currently much in vogue. These may be purchased in the pet stores. The most popular types are almost triangular in cross-section with a sliding acrylic front panel. They are very hygienic as they are cast in one piece and the internal angles are rounded, leaving no cracks or crevices for parasites to lurk. Ventilation panels are set into the top, the back, and/or the ends. With such cages the heat source, if used, will normally be placed outside the terrarium (heating pads or tapes).

Timber Terraria

Used in conditions of high humidity, timber will deteriorate by rotting, but it can be a useful material for constructing terraria not subject to excess moisture; desert conditions for example. The handyman can readily make a presentable display which is basically a wooden box with a glass or clear plastic viewing panel at the front. The type of timber used can vary from solid hardwood or pine planking to plywood of varying thicknesses. Chipboard, fiberboard, hardboard, and other manufactured timbers are not generally recommended for terrarium construction due to their moisture-absorbing qualities, though they do have their uses if adequately protected. The easiest material to use is probably 12 mm (half-inch) thick, exterior quality plywood, which may be glued and tacked together without a frame. The exterior of a plywood terrarium can be made to look most attractive by staining and varnishing to match or to contrast with the other furniture in the room where you wish to have your display. The interior can also be varnished or painted in colors to compliment the display in the terrarium. You must always allow several days for paint, varnish, etc., to thoroughly dry out before adding any animals to the cages.

The front viewing panel, which may also double as an access door, may be mounted on a hinged frame. It is best to have the hinges on the bottom and let the door swing outward and downward; this helps stop snakes from escaping, and also allows you to let the door hang right down during maintenance operations rather than having to prop it open. Alternatively, the glass may be slid into a groove, horizontally or vertically. In larger terraria it may be necessary to have two or more doors which may be separated by one or more fixed windows. Sometimes it is possible to get cheap, aluminium-framed sliding windows from a demolisher's yard; these make an ideal front to a

If you don't mind going to the time and trouble, you can house tree-dwelling snakes in a hand-built setup like this one.

terrarium which can be built to fit the window.

Ventilation holes should be drilled in the upper third of each end of the terrarium and also in the roof. Remember to ensure that the holes are much smaller in diameter than the smallest snake you intend to keep! To be doubly secure against escapes, the ventilation holes may be covered with a sheet of perforated zinc or wire gauze. Where very small, harmless species are being kept, the ventilation holes can be covered with PVC mesh (insect screening is ideal). The edges of the gauze should be secured and tidied up with a narrow framework made from half-round beading. Provision should be made for concealment of the heating and lighting apparatus, which will be discussed later. It may be possible to purchase a plastic base-tray (of the type used for cat-litter or something similar), and the terrarium could be built around the dimensions of this so that you will have a sliding floor tray. Such a tray is hygienic, will hold the substrate material, will prevent moisture damaging the wooden floor of the cage, and is easy to remove for routine maintenance or cleaning. The tray must be tight-fitting, however, preferably with retaining lips just above the edges of it to prevent the snakes forcing their way beneath it.

Built-in Terraria

The most satisfying and permanent kind of terrarium is one which is made from substantial materials, such as concrete blocks, clay bricks, or timber. It may be built into an alcove or be free-standing in a living area, conservatory, greenhouse, or spare room. The advantages of built-in terraria, which are particularly suitable for larger species such as boids, are that they can be constructed to match the internal decoration or form a focal point. A permanent, drainable, concrete pond can be constructed in the floor, while artificial rocks can incorporate controllable refuges for the snakes. Plant troughs can be built into the decor both in and outside the terrarium. Though there is virtually no limit to the designs which can be used, such a project is a major one and should be planned very carefully from the outset. Be sure to obtain advice from the appropriate tradesmen for any aspect of the building you are unsure about and remember that a brick and cement structure will be extremely heavy.

Terrarium Heating

There are several different kinds of heating apparatus, and which ones you use will depend on the size of the cages. For a number of terraria housing snakes from a similar climate it is usually sufficient to control the temperature of the room itself. Such a system is ideal for breeding-rooms in which larger numbers of small species or juveniles are kept in small containers. Where species with differing climatic requirements are kept, however, each terrarium should have its own individual heating system.

Whatever method is used, the apparatus must be installed and tested for accuracy before the decoration is completed and reptiles are introduced to the cage. Keep a thermometer mounted in each one so that you can easily see when anything is wrong with the heating system. As snakes seek out their preferred body temperature by thermoregulation, a range of temperatures should be provided. This is achieved by having the major heat source at one end of the cage so that various heat levels are available from one end to the other. A number of hiding places should be placed in various situations, and a snake will select the one in which it feels most comfortable. However, a snake may also move from

Above: Tetra Terrafauna **Hot Blocks** are substrate heaters which provide "belly heat" and help maintain ideal snake terrarium temperatures. **Below:** Another method of heating snakes is by use of something called a "spot heater" or "heat lamp," one of which can be seen suspended from the ceiling of this tank.

Some snake species need considerably more heat than others, and the boids are among them. Most boid species need at least 83°F/28°C of ambient heat, but some will require a temperature as high as 90°F/32°C and above. Shown is *Candoia aspera*.

one hiding place to another in order to adjust its temperature. Many species require a temperature reduction at night, and in most cases it is a matter of simply switching off the heat source; the room temperature of the average household is adequate for most species. However, where cages are kept in unheated rooms in cold winters, subdued nighttime heating will be necessary. The raising of the daytime temperature and its lowering at night should take place at the same times every day, but allow for seasonal changes. For example, temperatures will reduce slowly in the fall, reaching the minimum for the species in midwinter; the reverse will occur spring to mid-summer. The use of thermostats and time switches will be very useful and will ensure regularity of day/night temperatures which are not so easy to achieve manually, especially if you are frequently away.

Types of Heaters

Tungsten Lamps: In the past, the ordinary tungsten light globe was usually the only method of heating and lighting the terrarium. Such globes may still have their advantages when used in conjunction with other apparatus, as they are relatively inexpensive, emit light as well as heat, and come in various sizes. The required temperatures can be produced by experimenting with numbers of bulbs of various wattages. A useful means of nighttime heating is a bulb of low wattage colored blue or red to minimize the intensity of light.

Infra-red Lamps: Such lamps, of the type used in poultry hatcheries or by pig breeders, are very useful for larger terraria. Some manufacturers are now producing lamps especially for terrarium use. Such lamps have built-in reflectors which direct the heat into a limited area and are useful for heating up basking sites. Both red-light and white-light emitting lamps are available, and both have their functions (red lamps are useful at night in cold climates). The temperature of the glass bulbs in some of these lamps (usually available in 100-500 watt sizes) becomes very high, and care should be taken not to splash them with water when they are switched on as they may shatter. Basking spot temperatures can be adjusted simply by raising or lowering the lamp.

Ceramic Heaters: As these do not emit visible light, they are very useful for nighttime or supplementary heating. They may come in the shape of a bulb or a curved plate and may be fitted into a simple bulb socket. They release radiant heat; the bulb type in most directions, the plate-type in the direction in which it is aimed. The temperature emitted should be controlled by thermostat.

All forms of radiant heat lamps should be installed outside the terrarium with the heat directed through wire mesh, or they should be adequately encased with mesh to prevent the animals from coming into contact with them and getting burned. Never aim radiant heat directly onto terrarium plants or the plants will quickly desiccate and die.

Aquarium Heaters: Such thermostatically controlled glass-tube heaters are very useful to the snake keeper, especially if warmth and humidity are required. The heater is simply placed in the water bath and used as it would be in an aquarium. In smaller, humid terraria, where a large water bath is unnecessary, the heater may be placed in a concealed jar of water (which should be topped up at regular intervals). Some brands of these heaters can also be used out of water and may be placed in a special cavity of the artificial rockwork to heat up certain areas. Study the brands at your local pet shop to find out which are best suited to your needs.

Heating Cables, Tapes, and Pads: Electrical heating cables and tapes of the type used by horticulturists have their uses in the terrarium. A cable may be used to heat the substrate for burrowing and desert species, though they should be used in half the substrate only, giving the snakes a chance to seek out a cooler temperature if required. Tapes can be used for running along the bases of a number of small containers of the type used for rearing juvenile snakes. Aquarium heating pads will heat the terrarium from beneath.

Central Heating: Domestic central heating systems can be adapted for use in large terraria or snake rooms. When keeping several terraria containing species that require similar temperatures, this system can be used to heat the whole room. Alternatively, a system of pipes and radiators under cages can be installed in such a way that convection currents of warm air are directed through the terraria from below and out through the top. Thermostatically controlled valves and a bypass system will ensure individual temperature and time ranges for the terraria. Consult a heating engineer and discuss the options available. Though efficient and economical once installed, installation costs may be prohibitive, and such systems are perhaps only suitable for very large herp rooms or public reptile exhibitions.

Lighting

The practice of leaving snakes under strong light 24 hours a day is bad for them,

There are a now a wide variety of products being produced specifically for better herp keeping. Some maintain humidity, some help incubate eggs, and others, like the item shown here, help produce heat.

and daily light and dark periods (photoperiods) are necessary for all species, whether nocturnal or diurnal. Many diurnal species require natural sunlight or a good substitute.

Sunlight: As it is virtually impossible to artificially reproduce the intensity of natural sunlight, natural light sources should not be discounted. The terrarium may be placed near a window so that it receives a certain amount of sunlight each day. Remember, however, that the hot midday sun, especially in the summer, can quickly cause lethal overheating unless you provide an extremely efficient ventilation system. Morning or late afternoon sun is preferable for the confined interior of a terrarium, so an ideal situation would be near an east- or west-facing window. Natural sunlight through glass is of reduced quality, so part of the terrarium should be made of mesh or gauze. A portable terrarium may be placed on the veranda or in the garden on suitable days, so that natural sunlight reaches the reptiles.

Broad-spectrum Lighting: Broad-spectrum fluorescent tubes are specially designed to emit light from the blue end of the spectrum, the part of natural daylight which is most important to plants and animals. These include small but sufficient quantities of ultraviolet rays of an intensity unlikely to cause any damage to the reptiles' health. Originally produced for horticultural use, broad-spectrum fluorescent tubes are much safer to use than ultraviolet lamps, which now have limited use in the terrarium. In recent years, the use of such tubes has considerably improved the health and life expectancy of many species. Broad-spectrum lighting is essential in a planted terrarium if the plants as well as the animals are to thrive. Manufacturers are usually pleased to provide specification details of lamps available.

Halogen and Mercury Vapor Lighting: Quartz halogen and mercury vapor lamps both emit a greater intensity of light than most other readily available artificial light forms, and are ideal to use if lush plant growth is required. Such lamps are extremely powerful and may have wattages of 500 or more. They may emit a high level of radiant heat temperatures, so they are unsuitable for small terraria unless suspended well outside the cage.

Humidity
This refers to the amount of moisture in the atmosphere and is a major environmental factor in all habitats. For obvious reasons, snakes which are native to extreme low-humidity environments will not survive for long in a tropical rain forest terrarium. Conversely, it would be foolish to expect a creature native to damp woodland to survive for long in a desert terrarium. In general, it is probably easier to keep desert creatures than those from humid environments as the forms of heating and lighting in the average terrarium automatically tend to create a dry environment.

It will be necessary to provide artificial humidity levels for those species which come from moist habitats, but remember that some habitats are humid only at certain times of the year. The creation of higher humidity levels in a terrarium is not difficult. In small planted terraria, regular mist spraying of the interior is all that is required, though this may have to be done several times a day during dry periods. Spraying is important to many species of small tree-snakes, which acquire their drinking water solely from droplets on the foliage.

Where there are relatively large water surface areas, as in aqua-terraria, natural evaporation of the water will create local atmospheric humidity, which will be

Desert-dwelling species, like the Lyre Snake, *Trimorphodon biscutatus* shown here, will not need much in the way of humidity, but should instead be kept in relatively dry surroundings.

Those species that occur in very moist environments, like the Northern Water Snakes, *Nerodia sipedon*, shown here, are probably best kept in something called a paludarium or aqua-terrarium, which, in short, is a setup consisting of more or less equal land and water bodies.

enhanced by the action of the animals entering and leaving the water. Additional water evaporation will occur if an aquarium heater is installed, especially if an aquarium aerator is also used. At the same time you could consider using an airlift filter to help keep the water clear. A small air pump and an airlift filter, which you can obtain from your pet or aquarium shop, are indeed invaluable items in the aqua-terrarium or humid terrarium. They can be used for increasing the humidity, raising the temperature of the airspace, ventilating the airspace, filtering, and oxygenating the water all at the same time.

Although a humid atmosphere may be important to many species, it is important that snakes always have dry surfaces on which to bask or rest. Hide boxes or cavities should be placed so that they do not retain moisture. A permanently wet environment can lead to unpleasant skin problems which could be fatal. Seasonal changes in humidity may have a bearing on the reproductive cycles of many species, so some attempt must be made to reproduce these periodic wet and dry conditions in the terrarium.

Ventilation

Unless the terrarium is adequately ventilated, excessive carbon dioxide will collect in the terrarium base and a permanent dampness of the substrate will occur, leading to the development of harmful microorganisms and unpleasant-looking molds. In such conditions, the animals will suffer stress, lose much of their resistance, and possibly succumb to disease. In basic terrarium construction, provision must be made for adequate ventilation holes or panels. Heaters in the terrarium will create convection currents that will cause continual air exchange.

Sometimes it is advisable to provide additional ventilation apparatus rather than to rely on convection currents. An aquarium aerator will provide extra ventilation in a small terrarium and the air outlet need not necessarily be immersed in water if a lower humidity is required. For terraria kept in stuffy living rooms, particularly if the owners are smokers, it is advisable to have the fresh air drawn in from outside. This can be passed over an underfloor heater before it enters the terrarium through a special ventilator.

Furnishing the Terrarium

Some people favor the simple terrarium setup in which a sheet of absorbent paper, a dish of water, a hiding box, and a brick as a shedding aid are provided. This is indeed practical, especially where numbers of snakes are being kept for breeding purposes; the snakes do not seem to mind as long as they have the correct environmental conditions.

Most snake keepers, however, like to have at least one decorative terrarium in which their artistic talents can be used to produce a mini-habitat with landscaping and plants.

Floor Coverings: In the terrarium for small desert or semi-desert species, coarse sand or a mixture of peat (sterilized potting mixture makes a good substitute for peat) and sand can be used as a substrate material. Extra fine sand should never be used, as this will stick to the snakes' skin and could cake between the scales (especially after the reptiles have crawled through water), causing shedding problems later. For burrowing or semi-burrowing snakes the mixture should be relatively deep.

For larger snakes and humid terraria, the most hygienic decorative substrate material to use is gravel. The size of the gravel should be selected to suit the size of the inmates: 0.5 cm ($\frac{1}{4}$ in) gravel for small snakes, to 2.5 cm (1 in) or more for large pythons. Gravel should be thoroughly washed by hosing through with a strong jet of water, then dried out before use. It is always advisable to have a spare stock of gravel to use at cleaning time (which should be about once per month).

Rocks: Natural or artificial rocks can be used in the terrarium for decoration and to provide basking areas, hiding places, and aids to shedding. If you live in a big city where natural rocks are difficult to come by, you may be able to purchase suitable ones from aquarists' or gardeners' suppliers; however, it is much more fun to go out and hunt for your free rocks on location. Whether these are jagged rocks or large weathered pebbles is a matter of taste, but for the most attractive aesthetic effect it is best to stick to a single variety in each terrarium. Rocks should be arranged in the terrarium in such a way that they cannot fall down and injure the reptiles. In large terraria, where a great number of rocks are used, it is best to cement these together leaving only controllable hiding places, or great problems will ensue when the snakes need to be removed. With a little practice, natural rocks can be cemented together to produce a natural looking rock face. Judicious use of cement-coloring material can almost make the joints invisible.

Lightweight artificial rocks made from fiberglass or cement-covered plastic foam have proved very useful. Apart from the obvious advantage of a much lighter weight, they can be made to any size or shape and fitted exactly into corners so that there are no uncontrollable hiding places left for elusive snakes or parasites.

Tree Branches: A tree branch in the terrarium is attractive to look at and will provide exercise and a shedding aid for most snakes, especially arboreal and active species. A bleached, gnarled branch or root can look most attractive in the desert terrarium. Driftwood from the seashore or trapped in snags in rivers is often highly suitable as it will be weathered, smoothed down, and bleached by the actions of water, sand, and sunlight. Branches with an interesting shape could be lopped directly from the tree, but be sure you have permission if it is not your tree! Going out looking for bizarre branches can be almost as exciting as snake catching! All wood collected for the terrarium should be thoroughly scalded, scrubbed, washed, and dried before use. Hollow logs and branches are useful as hiding places for some species, but be sure that you have access; it can sometimes be very difficult to remove a stubborn snake that has jammed itself into a hollow log. With a little carpentry it is possible to make controllable hiding places.

Hollow logs can be used to grow epiphytic plants and form an attractive display in the rainforest terrarium. Additionally, dead branches can be used to support creeping plants as well as provide climbing frames for arboreal snakes. Large climbing snakes, such as boids, will require very strong, robust branches. All branches should be fixed securely in the terrarium so that there is no

Snakes' quarters need not be set up elaborately in order to be functional. Some newspaper, a water bowl, and a dish with a hole cut into it are all this albino Gopher Snake, *Pituophis catenifer*, will need.

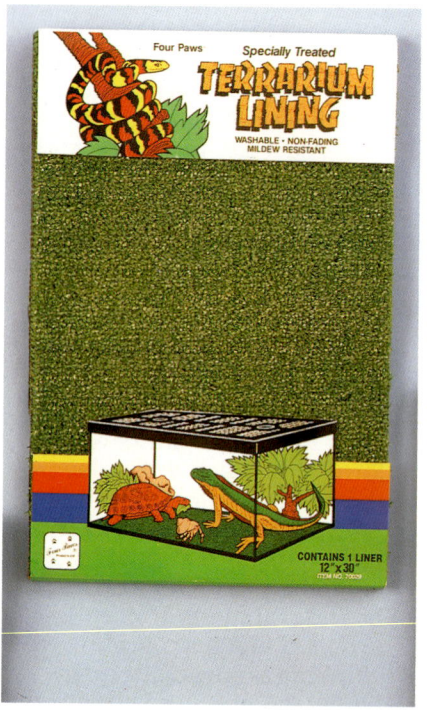

Fours Paws Terrarium Linings are fully washable and mildew resistant. The thick grass of the linings provides a happy and healthy environment for small animals.

danger of them collapsing and causing possible injury to the snakes. Corkwood bark of the type used by florists is a useful item and can be utilized as an attractive wall covering in the terrarium, to conceal plant pots, or to provide hiding places for the snakes.

Plants in the Terrarium

Plants are often more difficult to keep in good health in the terrarium than the animals and are not strictly necessary for the snakes. However, there is no doubt that a carefully planted and maintained terrarium is well worth the effort. It must be remembered that plants also have environmental requirements, and the types of plants used should be compatible with the environment you are creating for your snakes. A good book on exotic house plants will give you some tips on plant care. It is a good idea to have two sets of plants so that they can be changed over at regular intervals for recuperation. The plants should preferably be left in their pots to facilitate removal; the pots can be concealed behind logs or rocks.

It would be a complete waste of time trying to maintain living plants in a cage containing boids or other large snakes; they would soon be crushed and killed by the weight of the reptiles moving about. In such cases, artificial (plastic) plants should be used, or no plants at all. You can compensate by having a display of plants outside the terrarium.

Facing Page: Arboreal (tree-dwelling) snakes are often very difficult to house. They require much vertical space and implements on which to climb.

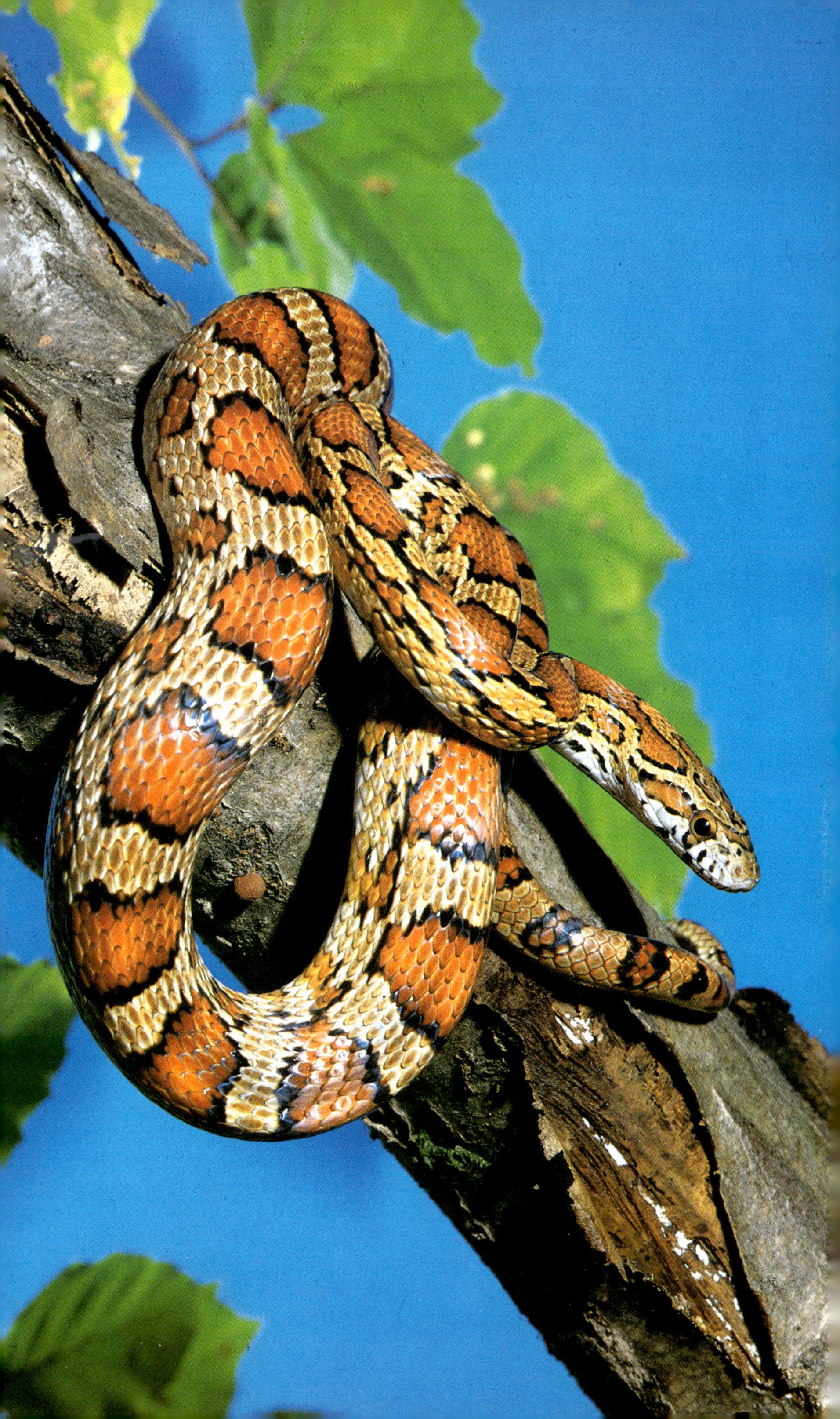

Nutrition

Some snakes have highly specialized diets and are thus best not kept by the beginning keeper. This Eastern Hognose Snake, *Heterodon platirhinos,* for example, feeds only on toads.

To maintain the bodily functions at an efficient level and thus remain in the best of health, a balanced diet containing the correct variety and quantity of constituents is required by all animals. Such a diet contains proteins, carbohydrates, fats, vitamins, minerals, and water in varying quantities.

As snakes usually feed on whole prey animals, they receive all of the basic dietary constituents. As long as the prey itself is healthy, there should be no need for any extra dietary supplements. An exception is those snakes (garter snakes for example) which can be trained to take strips of dead fish. This should only be used as a standby food when other food is not available, and should be used sparingly and dusted with a vitamin/mineral supplement. Freshwater fish should be given in preference to marine varieties, as the latter are more likely to contain the enzyme thiaminase which breaks down vitamin B1 in the body and will cause a deficiency of this vitamin.

Fresh water should, of course, always be available.

Some small tree snakes drink only from water droplets on the foliage, so daily misting of the terrarium is important.

Food Items

The choice of captive snake species may be influenced by what they eat. Species which specialize in feeding on certain items like frogs, geckos, or other snakes will often refuse to eat more easily available items. Unless a steady supply of special foods is available for particular species, then it is perhaps best not to keep those species in the first place. There is a great choice of other species which will take more readily available items.

General Invertebrate Foods: Small snake species and juveniles of many larger ones will feed mainly on invertebrates. In many cases, a variety of invertebrate food is important, and the best way to provide this is to collect insects, spiders, slugs, snails, earthworms, crustaceans, etc., from the wild. Termite nests are an invaluable source of food for some small burrowing snakes, and pieces of nests containing large numbers of the insects can be placed in the terrarium. In the temperate winter such food items become scarce, so one will have to rely on the limited number of feed species available from suppliers. These include crickets, locusts, and mealworms.

Crickets: Crickets are bred commercially, the most usual species being the house cricket, *Acheta domestica*. They are a good food supply for insectivorous snakes and are not difficult to breed. They may be kept in a small container such as a plastic aquarium, with rolls of corrugated cardboard or similar as hiding places. A shallow dish with a piece of sodden cotton wool will provide moisture for the insects to drink. Do not furnish an open water dish or the insects will drown. The crickets can be fed mainly on bran, supplemented with a little greenfood, carrots, or fruit. Hygiene is important in the culture, so uneaten food should be replaced with fresh food at frequent intervals.

Maintain the temperature of the culture in the range 26-30°C, which should allow the crickets to breed readily. They will probably lay their eggs in the damp cotton wool, so this should be checked regularly. If eggs are present, the cotton wool should be moved to a smaller container for hatching. The instars and adults will provide insects of progressive sizes.

Locusts: The migratory locust, *Schistocerca migratoria*, and other species are bred commercially. These are all highly nutritious foods for insectivorous snakes. Locusts require a well-ventilated cage and a temperature range of 26° C (night) to 33° C (day). They may be fed with bran, plus a daily supply of fresh grass stalks standing in a jar of water, the area round the rim being packed with cotton wool to prevent the insects from drowning. The moist cotton wool will also provide drinking water. Locusts are relatively more difficult to breed than crickets so it is perhaps best to purchase the occasional batch rather than attempt the difficult and time-consuming task of breeding them yourself.

Mealworms: These are possibly the oldest type of livefood bred commercially for many kinds of pet animals and can be useful to feed to very small insectivorous snakes. They are initially rather expensive to buy but are quite easy to breed from a purchased culture. The larvae of a kind of flour beetle, *Tenebrio molitor*, mealworms grow to about 2.5 cm (1 in) in length. They are generally best used only as a supplement to a varied wild-caught diet. Mealworms should be kept at room temperature in shallow trays with about 5 cm (2 in) depth of bran, crushed oats, or wheat, covered with a piece of sacking. One or two pieces of raw potato, carrot, or apple placed on the sacking and renewed as necessary will allow the mealworms to collect moisture. Some of the mealworms should be allowed to pupate and mature into adult beetles to ensure an ongoing supply of various sized worms as well as more pupae and beetles, all of which will be eaten by some snakes, especially small burrowing species. At intervals of two or three months, a new culture should be started with fresh bran. An alternative to breeding mealworms is to purchase a small amount as and when they are required.

Earthworms: Most earthworm species are a nutritious food for many small snakes. Garter snakes (*Thamnophis*), for example, will take earthworms almost as a staple diet, but fish should also be given occasionally. Earthworms are found almost everywhere and are common in most suburban gardens, especially where organic manure and mulch are frequently used (though worms from compost or manure heaps should be purged by keeping them in a mixture of moist bran for 24 hours before being fed to the snakes). A kind of worm "collection station" can be made by placing a layer of moist dead leaves, straw, or hay about 3-5 cm (1-2 in) thick in a secluded corner of the garden and covering it with a piece of sacking. Keep this damp, but not saturated, by regular hosing

Fish are a common item in many snake diets. Most feeder varieties can be obtained at your local pet store, or, if you're truly ambitious, you can set up traps and catch your own. The snake shown is a Northern Water Snake, *Nerodia sipedon.*

Pinkie mice are not only nutritionally complete but easy to obtain in both live and frozen form. A further consideration is breeding them yourself, which is also a fairly simple process. The snake shown is *Calabaria reinhardti*.

if necessary. Earthworms will congregate in the damp area to feed on the dead leaves and they may be collected daily for a period of 1-2 weeks by lifting the sack and sifting through the leaves. As soon as one area becomes sparse in worms (do not over-collect), a new station can be set up in another part of the garden. Do not collect again from the original spot until at least 12 months later. Live earthworms may be stored in containers of a moist (but not wet) mixture of potting compost and dead leaves.

Fish: Some aquatic or semi-aquatic snakes feed almost exclusively on fish. Some can be trained to take dead fish or even strips of fish meat, but it is best to give live, freshwater fish wherever possible. Due to the presence of thiaminase in marine fish, these should be given only to sea snakes. Readily-bred aquarium fishes such as guppies (*Poecilia reticulata*) or goldfish are a good standby item for small fish-eating snakes. Even frog- or gecko-eaters such as *Ahaetulla* or *Chrysopelea* can be persuaded to take fish if they are offered in a shallow container of water so that the fish flip about. Local small freshwater fish can be netted in rivers, streams, or lakes, but check local laws and catch fish in moderation.

Another good source of supply is the commercial freshwater fish farm where fish can usually be purchased at various ages from tiny fry to adults, and sizes suitable for most fish-eating snakes can be obtained. As some fish, such as trout, need to live in cool, highly oxygenated waters, it is advisable to buy only enough each time for the immediate requirements of your snakes.

Amphibians: Some species of snake may feed almost exclusively on frogs or salamanders, or maybe a mixture of both. Clumps of frog or toad eggs may be collected in the breeding season and allowed to hatch into tadpoles. These will be eagerly taken by many small aquatic snakes and can also be reared in aerated, planted aquarium tanks until they metamorphose into young frogs. The froglets can be further grown by feeding them on small insects so that you have various sizes for various sized snakes. The commercially bred African Clawed Frog, *Xenopus laevis*, is a very useful species as it can be bred all the year round so that a continuous supply of tadpoles and frogs of various sizes is available. The Axolotl, *Ambystoma mexicanum*, is a useful aquatic salamander that also breeds readily in captivity.

Reptiles: Many snake species feed partially or exclusively on other reptiles in the wild. Some arboreal snakes may feed on chameleons or geckos but will usually readily convert to skinks or anoles, for example. Many will even take small, live fish as a substitute. Useful food lizards for small snakes are the cosmopolitan House Geckos, *Hemidactylus frenatus* and *H. turcicus*, the latter of which has become feral in the Gulf States of the USA. A group of these geckos kept free in a warm, humid room or greenhouse (temperature 26-30° C) and given a supply of insects (crickets, mealworms, flies, etc.) to feed on will provide a self-sufficient supply of feed lizards after a period of time.

Snake-eating snakes will often steadfastly refuse to take any substitute voluntarily. As a regular supply of snakes for snake-eaters is not usually available, force-feeding of substitute foods is often required. The King Cobra, *Ophiophagus hannah*, is a typical example of a snake-eater which will usually refuse any other offerings, though force-fed specimens have lived for many years. Unless you have the time and patience for force-feeding, the most satisfactory solution is probably not to keep snake-eaters at all.

Birds: When we arrive at the subject of feeding birds and mammals to snakes we are touching on a controversial one. Many people are totally against others keeping captive snakes because they feel sorry for the animals with which they must be fed. Most snakes, however, will take dead birds or mammals after a period of training, and one should make it a policy to feed dead prey whenever possible. The most widely used food items for bird-eating snakes are day-old chicks of domestic poultry. These are inexpensive and can be obtained alive or dead from hatcheries, which usually have large numbers of rejects. Some specialist companies supply large quantities of deep-frozen chicks that can be thoroughly thawed out and used as necessary. It is useful to keep a few live birds and grow them up to various sizes, depending on the size of the snakes you have to feed. However, bear in mind that zoning laws in some areas prohibit the keeping of live chickens and ducks (even those a few days old). Other useful domestic birds (both chicks and adults) for snakes of various sizes include quail (the chicks of which are very small and suitable for small snakes; quail eggs are also good for egg-eating snakes), pigeons, ducks, geese, turkeys, and guinea fowl. Adult birds of the larger varieties are of course suitable only for the larger boids. The feeding of wild

Birds and their eggs are two common dietary elements among snakes. Above, a Yellow Rat Snake, *Elaphe obsoleta quadrivittata,* attacks a hatchling in its nest. Below, an African Egg-eating Snake, *Dasypeltis scabra,* has just swallowed a quail egg.

Rodents are perhaps the most well-known snake food. There are a vast number of snakes that will accept them. For keepers, rats and mice are perhaps the easiest to obtain. Here a Russell's Viper, *Vipera russelli,* is seen taking a large rat.

birds to snakes should be done with the greatest of caution. Birds which have been killed by traffic and are not too badly damaged or tainted may be used occasionally. Otherwise, only use those birds which are classified as pests and are not protected by law; these may include house sparrows and common starlings, but be sure you are aware of local laws before attempting to catch any (by humane trapping or netting). For obvious reasons, any animals which have been poisoned or shot with shotgun pellets should never be used.

Mammals: The main mammals used to feed captive snakes are mice and rats of the kind bred in laboratories. Some snake keepers like to breed these animals and ensure a steady supply; others may purchase them as they are required, thus saving the extra time of maintaining the breeding groups, which can often be more time consuming than the reptiles themselves. Like day-old chicks, mice and rats may be available deep frozen from commercial suppliers. Very young mice, sold as "pinkies," are often the best food for small constricting snakes. Other mammals useful for larger snakes include gerbils, gophers, hamsters, guinea pigs, and rabbits, which should always be humanely killed before feeding.

Needless to say, all animals kept purely as prey animals for snakes should receive kind and hygienic husbandry, which includes adequate shelter, clean bedding, a balanced diet, and fresh drinking water. They should be cleaned out at regular intervals and not be overcrowded. No animal should ever be neglected just because it is food for a snake!

Feeding Strategies

Frequency of feeding in captive snakes is often a matter of trial and error. Under wild conditions there is a natural balance between the amount of energy required in catching prey and the nutritional value of that prey. Thus, in captivity, species which feed readily can easily be overfed in the juvenile stage, and massive amounts of fat eventually diminish the functions of the internal organs, often leading to sterility and even premature death. It is also difficult to make any rules about amounts of food to be given at each meal, and a little experimentation will be necessary. It is perhaps better to err on the "too little" rather than on the "too much" side, but the following generalizations can be taken as guidelines.

Juvenile growing snakes should be fed about twice per week. Small- to medium-sized snakes can be offered food about once per week, while larger snakes can be fed substantial meals about once every two or three weeks. Any snake that is obviously becoming overweight should have its food intake reduced.

When two or more snakes are kept together in a cage, keep a close watch at feeding time as more than one snake may grab the same prey, resulting in injury or even one snake swallowing the other. Snakes do not feed during the shedding period, which may occur from three to eight times per annum depending on age and species. Always watch for signs of shedding and do not feed the snakes at these times. Introduction of live prey can be especially dangerous to the snake at a time when it is disabled by the shedding process. Indeed, careful watch should be kept when live prey is given and, if it is not overpowered and eaten in a few minutes, it should be removed from the terrarium and tried again a day or two later. Live mice and rats have been known to gnaw into the body and severely damage an uninterested or temporarily disabled snake at night. Most snakes can be persuaded to take dead prey, particularly if it is moved about in front of their snouts; it is best to do this with a stick to avoid getting bitten. Dead food animals may be left in the terrarium overnight, but should be removed and discarded the following morning if uneaten. Dead prey will putrefy very quickly in the warm environs of the terrarium.

The natural food of all snakes is whole prey animals, which they almost always capture alive. The keratin, fur, feather, and bones of the prey animals are all an important part of the snake's diet and provide roughage as well as valuable minerals for bone building and organic functions. With the exception of some fish and invertebrate feeders, snakes will not normally require any additional supplements to their diet, as whole prey animals are a balanced diet in themselves. However, snakes being given pieces of meat, fish, or invertebrates should have a vitamin/mineral supplement dusted on the food.

Occasionally a snake will steadfastly refuse all food offered to it. After ensuring you have given the correct food for the species, have tried various kinds of foods, and have offered it dead and alive, during the day and at night, and still had no success, it will be necessary to resort to force-feeding. Before doing this, however, ensure that the reptile is not suffering from any disease or condition which may contribute to it losing its appetite. Mouth rot (necrotic stomatitis), shedding difficulties, internal or external parasites, and many other conditions may affect the appetite, so treatment for such conditions should precede or accompany force-feeding.

There are two methods of force-feeding. The first is to take a whole dead prey animal of suitable size. Opening the snake's mouth by gently pulling on the loose skin below the jaw, the prey animal is pushed head-first into the buccal cavity. Sometimes the snake will then start swallowing the prey of its own accord, in which case it can be released. If not, it will be necessary to work the prey down into the gullet and massage it into the stomach. The handle of a wooden spoon (lubricated with animal fat, glycerin, or mineral oil) or something similar can also be used to gently push the prey home.

If the first method fails, the second method of force-feeding is to liquefy the dead prey animal in a food processor (a little water may be added to make a slurry), then place it in a large syringe with a smooth-ended rubber or plastic tube of suitable diameter and length pushed over the nozzle. The tube is lubricated with animal fat, glycerin, or mineral oil and pushed into the labial notch of the snake's mouth and then down into the stomach (approximately one-third the length of the snake). The

Many newborn snakes will only be able to take pinkie mice.

contents of the syringe are then simply squeezed out. In order to get all of the food out of the tube and into the stomach, an area above the slurry in the syringe can be filled with plain water. Problematic feeders can often be maintained for long periods using these methods, which seem to do them little harm. However, this type of feeding is on the whole very unnatural and not recommended for extended periods of time.

Even snakes that normally won't bite may have a change in temperament if they are hungry enough. This is one good reason to always make sure your pets are well-fed.

General Care

Before making any rash decisions, the aspiring snake keeper should first give great consideration to the responsibilities involved. Remember that an animal confined to a cage cannot care for itself, and you are responsible for its welfare and all its requirements. Before you obtain any specimens, ensure that you have the time, patience, and ongoing enthusiasm to care for them properly. The possession and care of snakes should be a great pleasure and relaxation to the keeper, and anybody who is unsure about this aspect of snake keeping should never start!

A further point to be considered is, while the potential snake keeper may be enthusiastic about their prospective hobby, others, including close members of their family, may not share their enthusiasm. People who like snakes tend to have a "feel" for them, a quality not possessed by most other people. Once those around you have overcome your eccentricity, you can usually go ahead; but remember that while some people can be converted to like snakes and others can learn to tolerate them provided they are kept locked up, there are a few who will not view them with the slightest favor under any circumstances. The herpetologist should view such people with muted sympathy and on no account should he use his snakes, even in fun, to instill further fear.

A few snakes in the home need not be overly time-consuming in their maintenance. Once the basic terrarium has been set up, four or five hours per week should be adequate for routine husbandry and care.

Usually only a small percentage of snake species are available to the hobbyist, and what is available often depends on such factors as the season, where you live, and the conservation status of the snakes in question. Many countries have specialist herptile dealers, and many pet shops will have more than a few snake species. Avoid buying from premises which are obviously dirty and untidy, though fortunately these are now few and far between. Modern animal welfare education and legislation have produced professional dealers whose premises are kept absolutely spotless and who offer for sale only animals in the best of health. Such animals are always displayed in clean, attractive display terraria and the dealer is always pleased to give specific

Some hobbyists turn to the wild in order to find specimens for their collection, but this practice is slowly becoming taboo.

advice regarding the habits and care of particular species.

A dealer is unlikely to guarantee the health or life of an apparently healthy specimen once it has been taken from their premises (though some will guarantee live delivery if animals are dispatched by contract transport), so examine animals carefully before purchase. Look for signs of ectoparasites both on the animals and in the display cages. Ask for a trial handling. If it is your first time, the dealer will be pleased to instruct you. Choose only those snakes which are plump, sleek, and have an unblemished appearance devoid of any unshed patches of skin. Ensure that the reptile is alert and that its eyes are bright and clear. Examine its mouth and vent for signs of inflammation which could indicate disease. Ask if the snake is feeding and what it is eating. It will be a great help if you know from the outset what the snake has already been eating. It is often suggested that you actually witness the snake feeding, but this is not always possible.

A good method of acquiring stock is to

purchase it from other hobbyists who have bred their snakes or who wish to dispose of excess stock. The best way to find out who may have excess stock for sale is to join a local herpetological group if there is one in your area. By regularly attending meetings you will gradually learn who is keeping and breeding what species, and not only will you be able to obtain good stock but you will have first-hand advice on how to look after it. Advantages of obtaining home-bred stock (as opposed to wild-collected stock) are that it will probably settle more readily into its new terrarium and is less likely to be diseased or infested with parasites. Another good point about home-bred stock is that exchanges can be arranged so you can either diversify blood lines or obtain new species.

A third method of obtaining specimens is collection from the wild. This method should be pursued with great caution as many countries now have a total ban on the collection of wild specimens, though this doesn't stop the locals from hacking to death any creature that even vaguely resembles a snake. In some countries certain species may be protected in some

Litter piles in wooded locales are often the most productive sites for snakes.

areas but not in others. If collecting is allowed, it is ethical (and good conservation practice) to collect only what one requires, so you must overcome the temptation to overcollect when species are abundant. One advantage of collection from the wild is that one can see exactly the type of habitat in which the snake lives and prepare the terrarium accordingly. Before attempting to collect from the wild, always be aware of all national and local legislation pertaining to the area in question.

Snakes are best transported in individual cloth bags, these being placed in stout cardboard or wooden boxes. When packed in bags, snakes seem to enjoy the feel of the cloth close to their bodies and quickly become calm. They can easily breathe through the pores in the cloth as well. Close-woven material like linen or calico is ideal for making snake bags, and if you have someone in the family with a sewing machine you can soon make up a number of them quite cheaply. Rice or flour sacks often have their uses, and in emergency situations you can always borrow a pillowcase. It is convenient to have a tie strip sewn near the neck of the bag so that it can easily be tied shut. Before use, bags should be thoroughly inspected for holes.

For regular snake transporting it is wise to invest in one or more carrying boxes. These are best made of stout timber with ventilation holes drilled in the lid. The lid can also have a number of cup hooks screwed into it on the inside so bags can be tied up individually; alternatively you can have several compartments. This is for safety's sake, so you do not have to put your hand among a pile of bags containing venomous snakes. Remember that snakes can, and will, bite through the cloth, so always pick up the bag near the neck, above the tied tape.

It is best to collect specimens yourself from dealers rather than have them sent by some form of public transport. The dealer will usually supply bags and/or packing boxes. In cold climates, the boxes containing the bags should preferably be lined with an insulating material such as plastic foam, just leaving a few ventilation holes in the lid. Unlike birds and mammals, a snake's oxygen requirements are minimal in cold weather, so it is better to insulate rather than over-ventilate during transport.

Quarantine

Newly-acquired snakes arriving into a collection should always undergo a period of quarantine to reduce the possibility of disease being introduced into existing stocks. Simply-

Snakes of any species can be unpredictable in temperament, and in the case of larger specimens it's best not to risk being bitten. Hold such snakes in the same manner this Horseshoe Racer, *Coluber hippocrepis*, is being held here.

furnished quarantine cages should preferably be kept in a different room from the main collection. Keep new stock under close observation for not less than 28 days. If all is well after that time they may be introduced to the permanent collection. If a disease breaks out while the animals are in quarantine, they should receive veterinary treatment and remain isolated until 28 days after a complete cure has been effected.

Inspection and Handling

New arrivals must be thoroughly inspected for signs of diseases and parasites, and if anything is discovered they should receive the appropriate treatment. In any case, it is advisable to inspect the reptiles thoroughly on a monthly basis to ensure they are remaining healthy. Many snakes, especially some of the larger boids and colubrids, soon accept being handled on a regular basis; others may remain nervous and aggressive, and there it is a possibility that overhandling in all snakes can interfere with breeding cycles. Therefore, unless a snake is to be kept as a non-breeding pet it should be handled as little as possible. Once a pet snake is tame, however, it should be handled frequently to keep it tame. After becoming accustomed to the human smell and relating it to no dangers, most snakes will remain docile.

For obvious reasons, venomous snakes should never fall into the category of "pets." They should be handled as infrequently as possible, and preferably not at all.

We do not support or encourage the keeping of venomous snakes in any way.

Methods of handling vary with the size and the aggressiveness of the species or individuals, and can be divided into four groups.

Small Nonvenomous Snakes: Aggressive specimens up to, say, 50 cm (20 in) in length, should be grasped with the thumb and forefinger firmly, but gently, around the neck just behind the head, allowing the body to drape over the hand. Non-aggressive or tame specimens may be simply lifted up by placing the fingers under the body and allowing the snake to drape across the hand or twine between the fingers. Small children should always be closely supervised when handling snakes, as they tend to grip too tightly and could injure the reptiles. Do not hold any snake by the tail as this can cause spinal injuries. Bites from small non-venomous snakes are negligible as the teeth are so tiny they are unlikely to break the skin. The worst you could expect are very minor pinprick-like

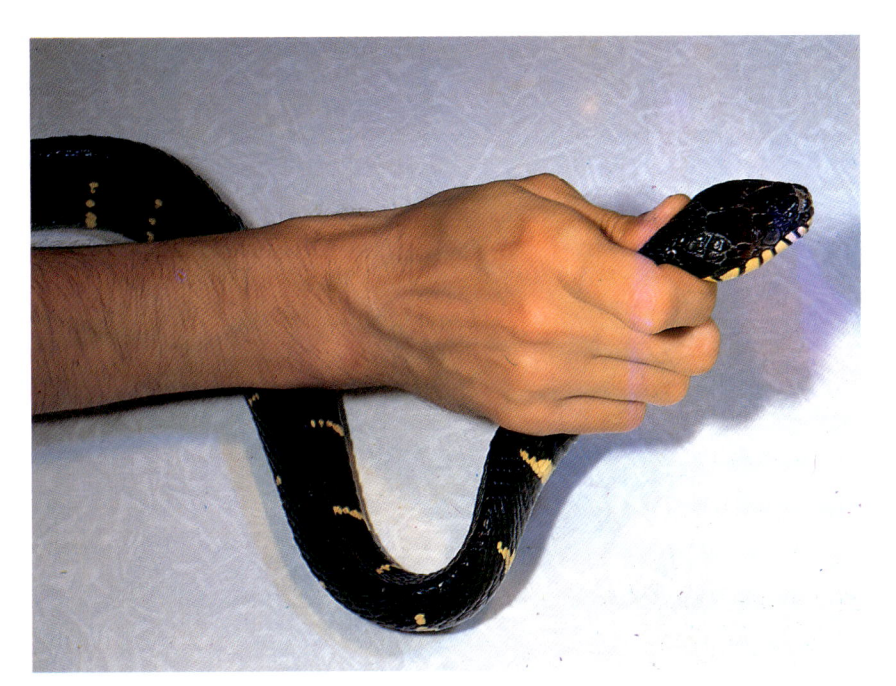

When handling any type of venomous species, caution is the priority. When grasping the snake with your hand, as shown above, make sure your grip is firm, and just behind the head. When immobilizing the snake beforehand (below) use something to pin the head down. The species shown is the Mangrove Snake, *Boiga dendrophila*.

653

punctures, which can be simply wiped with a mild antiseptic preparation.

Medium-sized Nonvenomous Snakes: These include snakes from 50-180 cm (20-72 in). Some, depending on the size of their teeth, are capable of giving deep, painful, or even lacerative bites which may require medical treatment. Aggressive or lively snakes should be secured by the neck just behind the head with one hand, while supporting the body with the other. Do not hold the snake by the neck with one hand and allow it to thrash its body about, as it will be likely to injure itself. A tame snake in this category can be simply lifted up about one-third and two-thirds down the body using both hands. It can then be draped over the arms and restrained occasionally as it crawls about.

Large Non-venomous Snakes: The larger colubrids and many boids in excess of 180 cm (72 in) are capable of giving deep, lacerated bites that may require medical attention. The constrictive powers of some of the larger snakes should also be respected— remember that people have been killed by large constricting snakes. Large, aggressive specimens should be handled by two or more people. One person grips the snake firmly by the neck, just behind the head, while the other(s) restrain its body. Many large boids become remarkably tame, especially if handled frequently from the juvenile stage. Large, docile snakes can be simply lifted by the body using both hands and draped around the arms and shoulders. Overalls or old clothes should be worn when handling nervous snakes since they have a habit of emptying the contents of their cloaca when restrained. The fecal discharges are often foul-smelling and persistent and can be corrosive to clothing unless quickly rinsed off.

Venomous Snakes: Venomous snakes are definitely in a category of their own when it comes to handling. All venomous snakes (including those rear-fanged species which are regarded as "mildly" poisonous) should be treated with the utmost respect. Venomous snakes should **never** be kept by beginners and certainly not by children. Any prospective handler of venomous species should first have had plenty of experience with aggressive non-venomous varieties in order to be familiar with the type of problems that may occur. When possible, it is highly recommended that a short apprenticeship is served with an experienced venomous snake handler before any specimens are acquired. The prospective keeper of venomous snakes should be thoroughly

654

familiar with all the rules, regulations, and safety procedures associated with venomous snakes and their venoms.

The viperids and crotalids are generally somewhat easier to handle than most of the elapids, though there are exceptions. Some of the heavier viperids and crotalids are fairly sluggish and cumbersome in general movement, though they can strike in a flash if necessary. These heavy snakes can be simply picked up with a snake-hook, which consists of a metal rod with a "T"- or "L"-shaped end. The hook end is simply slipped under the center of the snake's body and lifted up and, as the snakes do not like to fall, they will usually stay there until you put them down. With very heavy vipers it may be necessary to use two snake hooks or a combination of hook and grab. Ensure that the handle of the hook or grab is longer than two-thirds the length of the snake and always hold it well away from the body. When handling venomous snakes it is wise to have as few people as possible in the vicinity, and always ensure there is a responsible person within earshot in case of an accident. Elapids and rear-fanged colubrids should be handled with a snake-grab or clamp stick. This consists of a long rod with a handle and trigger-like grip at one end, and a cable-controlled tong at the other. Such instruments as fruit or litter pickers can be converted for snake handling. The tong should be cushioned with rubber tubing or foam rubber to prevent possible injury to the snakes. Special snake grabs are manufactured by some companies; information on these may be obtained through your herpetological society or your local zoo.

Venomous snakes should be grasped in the hand only for specific examination, medical treatment, or venom extraction (milking). The snake's head should first be pinned to the surface on which it is resting with the "T" end of a snake stick. Firm pressure must be used, but do not apply so much that the snake is injured. It requires a bit of practice before you get it right. Having secured the snake's head with the stick, the other hand is used to grasp the neck with the thumb and forefinger just behind the head at the angle of the jaws, ensuring that you have not left enough neck free for the snake to turn its head around and bite. The snake stick can then be placed to one side and the other hand used to restrain the snake's body. The technique of handling venomous snakes is one that can never be satisfactorily described in writing, and unfortunately the only way to gain experience is by doing it, preferably in the company of somebody already experienced. The appropriate antivenom should always be available, preferably with the local medical practitioner or at a local hospital where expert medical attention will be available in case of accidents.

Never be complacent when dealing with venomous snakes. A good herpetologist should never get bitten. After receiving a relatively mild, but at the same time extremely painful, bite in the thumb from a European Adder, *Vipera berus*, when I was a teenager, I became very cautious. But I have still managed to get myself into a few potentially fatal situations over the years and I will relate one of these stories just to show how easy it is to become complacent.

Once I had been working on a bank of three small, eye-level cages in a public reptile house. Two of the cages were empty, the third contained a Malayan Pit Viper, *Agkistrodon rhodostoma*. I placed my tools in one of the empty cages, locked it, and turned off the cage lights to go for my coffee break. On returning, I unlocked the cage containing the Viper by mistake. Not only did I do this, but I then stuck my head into the cage in order

Care must always be exercised with venomous snakes even when performing a simple function like changing a water bowl.

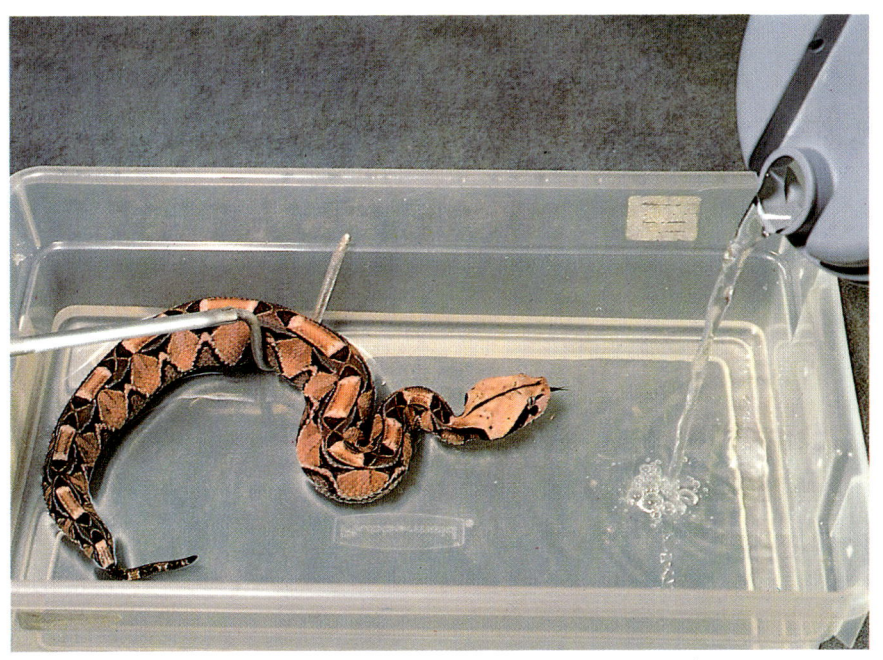

Before hibernating any snake, it is always a good idea to give it a few warm water baths. This helps completely empty out its system.

to look for my hammer. A sudden flash of white right in front of my eyes made me withraw my face quickly and slam the door shut as I realized what I had done. Yes, the snake had struck at me and the white flash had been the interior of its mouth, but fortunately it had only been a warning strike. It could just as easily have bitten me in the face, however, and my chances of survival would have been extremely slim, bearing in mind the tissue destructive qualities of the venom of this species and the closeness of the bite to the brain. You can imagine how I felt for several hours after the incident!

Hibernation

Snakes are relatively non-adaptive to conditions alien to them and, although a reptile from a temperate habitat may appear to thrive and feed in summer and winter under artificial heat, its life will be considerably shortened and its cyclic rhythms will be disturbed if it misses out on hibernation. Snakes kept in such conditions are unlikely to breed. Temperate to subtropical snakes require seasonal changes in temperature and photoperiod as well as a period of torpor during the winter months. In the wild, such periods occur when the temperature is inadequate to permit normal activity. Hibernation also plays an important part in seasonal reproductive activity, which normally takes place in the spring. Wild snakes usually hibernate deep in the ground in fissures, hollow roots, or burrows of various animals; some species congregate in large numbers in suitable hibernacula.

Hibernating a captive snake for several months of the year may seem a boring proposition. However, an adequate alternative to hibernation seems to be a short rest period at a much reduced temperature. This can be achieved by first ensuring that the reptile is adequately fed and has built up enough body to be able to withstand a prolonged period of fasting. Feeding is then stopped and the temperature in the terrarium is reduced progressively over a period of about 14 days. The terrarium is best kept in an unheated but frost-free room at this time. When the temperature is reduced to 4-6° C, the reptiles can be allowed to hibernate in the terrarium (under cover in their hiding places) or placed in a ventilated container loosely packed with dry sphagnum moss. The reptiles are kept at this low temperature for 4-8 weeks, after which the process is reversed: a gradual increase to 20-28° C over a period of 14 days, but still reducing the night temperature to around 15° C. Such treatment will usually trigger a natural breeding response from snakes which normally hibernate in the wild.

For species that normally occur in temperate regions, like this Eastern Milk Snake, *Lampropeltis triangulum triangulum*, hibernation is the essential first stage in the breeding season. Unhibernated females may still produce eggs, but chances are they will not be fertile.

Hygiene and Disease

Infected eyes of an imported snake, Blood Python, *Python curtus*.

Good hygienic practice is essential to minimize the risk of disease and retain our snake specimens in prime health. The most hygienic substrate on which to keep snakes is absorbent paper that can be easily changed each time it is soiled. If the reptiles are kept on other kinds of substrates, such as gravel, sand, or leaf litter, the fecal pellets can be scooped out as they appear, using a large spoon or a small, long-handled shovel. As such floor covering material is easy to replace, it does not matter if you remove some of it with the droppings. The feces of a healthy snake is normally a fairly solid pellet which leaves little mess behind once you have scooped it out. The cage should be given a more thorough cleaning out about once per month. The snakes may be removed to a spare cage or temporarily placed in a secure container (a clean trash can is ideal) during these chores. All substrate materials, decorations, water vessels, etc., should be removed from the cage either for trashing or thorough cleaning. The inside of the cage and any accessories should be scrubbed with warm, soapy water, followed by a weak solution of bleach, then finally swilled out with clean water before being dried and refurnished. If possible, allow natural sunlight to dry out the cage interior, as this in itself is a good additional disinfectant. Domestic disinfectants other than bleach or povidone-iodine should never be used.

Fresh drinking and/or bathing water should always be available. Many snake species seem to delight in defecating in the water bath (often just after it has been cleaned!), so cleanliness is essential. The glass viewing panels of the terrarium should be crystal clear at all times; nothing destroys the esthetic effect of a snake's tank like a viewing glass smeared with filth. All these regular chores may be a little time-consuming but they are well worth the effort.

Personal hygiene is also of utmost importance when dealing with captive animals. Although snakes are not notorious for passing diseases on to humans, there is no need for unnecessary risks. Always wash your hands thoroughly after handling the contents of each terrarium to avoid the possibility of spreading potential disease organisms. It would be wise to have overalls or old clothes specially for use when you are dealing with your snakes.

Diseases and Treatment

Snakes kept in optimum conditions and receiving hygienic management are unlikely to succumb to disease. New arrivals to a collection, from whatever source, should be quarantined for a period of anywhere from one to three weeks before introduction to existing stock. Any snakes which become sick should be isolated immediately, preferably in a special hospital cage with minimum furnishings. This is best kept in a separate room to the main collection.

All snakes are valuable possessions and deserve professional treatment. Do not attempt to diagnose and treat diseases without veterinary advice. Today's veterinarians are taking a more active interest in pets other than dogs, cats, and farm animals, which had hitherto been somewhat neglected. There are vets who show enormous interest in and knowledge of reptiles and their diseases and are willing to advise your local veterinarian through the appropriate veterinary associations. Additionally, it is important to be up to date with regular advances and changes in the various drugs used for the treatment of sick reptiles.

Terraria which contained diseased snakes should be thoroughly disinfected by using a strong sodium hypochlorite (bleach) solution, ensuring that all parts of the cage, its surrounding areas, utensils, and furnishings are treated. Dead snakes should be sent for an autopsy; certain institutions and pathologists may be prepared to perform

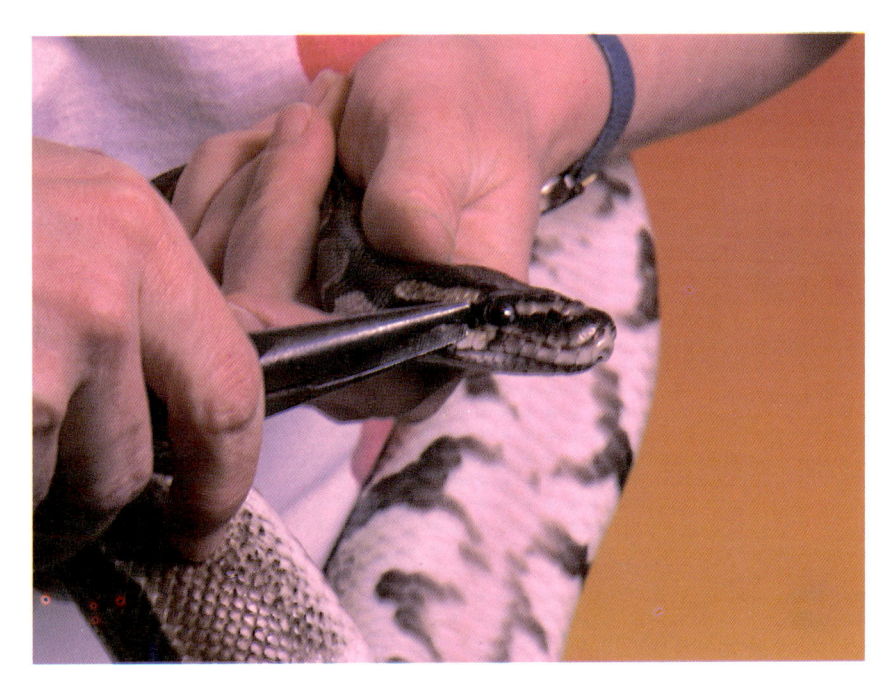

Shown here are examples of two of the most common ailments affecting captive snakes. Above, a tick is being removed from a Ball Python, *Python regius*. Below, a Boa Constrictor, *Boa constrictor*, suffers an advanced case of infectious stomatitis, otherwise known as mouth rot.

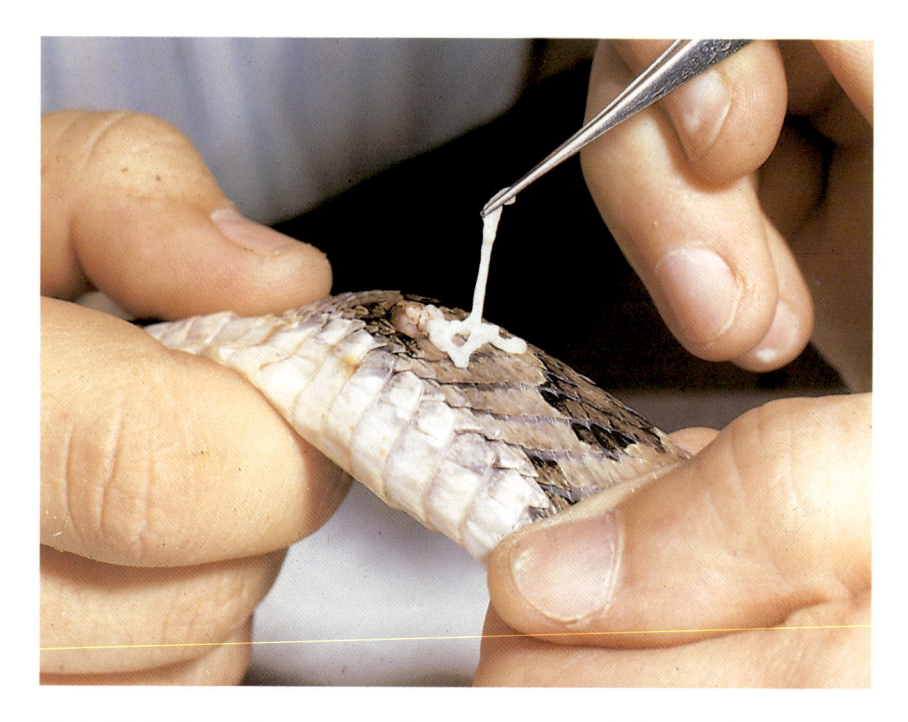

Most skin lesions will cause a snake great stress and thus should be treated immediately. Above, a viper is coated with a topical ointment, and below, a python shows the sutured wound that resulted from a fight with a cagemate.

such examinations free of charge or at a small fee in the interests of science. Your local zoo, college, or museum may be able to tell you where the snake may be sent for the autopsy. If a reptile dies from a known disease, or if it is impractical to send it for a post-mortem examination, it should be incinerated. A dead snake for examination should be placed in a plastic bag tied at the neck and then placed in another bag. If it is likely to be some time before it reaches the laboratory it should be injected with formalin in several parts of the body, including the thoracic and abdominal cavities and the fleshy part of the tail. This will delay the putrefaction and breaking down of the organs.

Although diseases should be relatively infrequent in captive snakes kept in optimum conditions, it is wise to be able to recognize the symptoms of the better known conditions and diseases, a brief summary of which follows.

Stress: Most people are surprised when I tell them that snakes can suffer from stress. Those most susceptible are snakes which have been recently captured from the wild; the trauma of being caught, transported, and placed in an alien environment can take a toll on the reptiles' health. Wounded snakes will also suffer stress, as will those subjected to suboptimum captive conditions. Stress reduces resistance to disease, allowing organisms that normally live in harmless symbiosis with the snakes to become suddenly pathogenic. Precautions should thus be taken to minimize stress as much as possible. Newly arrived specimens should be treated with the utmost care and respect and handled as little as possible. Sudden movements in front of the quarantine cage should be avoided; in fact, it would be wise to cover the glass viewing panel for a few days until the snake becomes accustomed to its new home.

Being very much creatures of habit, snakes soon learn exactly where the various basking areas, hiding places, water bowls, and other terrarium furnishings are situated. Once a snake is housed in its permanent terrarium the furnishings should remain in their original positions and not be moved around each time the cage is cleaned out, otherwise the snake will have to become accustomed to its new layout time after time, which could lead to a stressful situation.

Wounds: These may be suffered as a result of fighting, attempting to escape, striking against the terrarium glass, and so on. Over-enthusiastic capturing techniques and encounters with dogs or cats or the garden spade also contribute to wounds which will require treatment if the snake is to survive. Burns resulting from close encounters with heating apparatus will also require treatment. All such wounds are susceptible to infection, particularly in the case of newly captured specimens suffering from stress. Open lesions should be treated as soon as possible by bathing them with an antiseptic solution such as povidone-iodine and, if extensive, stitched by a veterinarian with an everting mattress suture. Healing of such wounds is often prolonged in snakes, and a regular course of antiseptic bathing should be carried out until visible healing begins. In some cases, antibiotic treatment may be required. Bactericidal antibiotics can be used in snakes for both prophylaxis and treatment.

Environmental Deficiencies: The best environmental conditions for a captive snake are those which are as similar as possible to those in its natural habitat. Should the temperatures be too low, the snake may refuse to feed or regurgitate recently consumed food. Protracted digestion may result in putrefaction of the prey in the snake's body and lead to food poisoning. If the temperature is too high, heat prostration with renal constipation will result in death. Overly damp surfaces or excessive humidity levels will result in one or more skin diseases in some species, while others may be adversely affected by situations which are too dry—respiratory or shedding problems including retained spectacles, for example. Photoperiod, light quality, and seasonal environmental variations can also affect the natural cyclical activities such as mating, shedding, and hibernation. The various life support systems in the terrarium should therefore be regularly checked and adjusted if necessary so that conditions remain optimum for the individual species being kept.

Nutritional Problems: Snakes which feed readily are unlikely to suffer from serious nutritional disturbances other than obesity resulting from eating too much. One exception concerns those snakes which may be fed on fish or pieces of fish meat. The flesh of many fish, but mainly saltwater species, contains the enzyme thiaminase, which destroys the B1 vitamin (thiamine) in the body and can result in a deficiency of this vitamin (hypovitaminosis B1) with symptoms of disorientation, loss of balance, and convulsions. If fish is heated to 80° C for about 5 minutes the thiaminase is destroyed and the balance returned.

The fish of course should be allowed to cool to room temperature before being fed to the snakes. Snakes that have starved for long periods for one reason or another will be emaciated and obviously deficient in vitamins and minerals. Once the snake begins to feed again, or is force-fed on a regular basis, the balance is restored. In severe cases, the veterinarian may recommend the addition of a vitamin/mineral supplement to the food to speed up the return of a diet balance.

Shedding Problems: A frequent problem in captive snakes is difficulty in shedding satisfactorily. Snakes cast off their old, worn skins at irregular intervals, the frequency and time depending on species, age, and seasonal influences, but usually three to eight times each year. Difficulty in shedding completely may occur if the relative humidity in the cage is too low or if the snake is suffering from an infestation of skin mites. In such cases, the skin will blister up and break off in pieces, allowing the lubricating fluid to dry out under the unshed pieces and causing them to stick tightly. Such unshed skin must be removed, as infection can quickly set in beneath, creating further health hazards. The affected reptile should be placed in a container of lukewarm water in such a way that its body is completely submerged but the head is above water. After allowing it to soak for a couple of hours, some of the loose skin should float away and the remainder can be gently peeled off with the fingers.

Occasionally an old eye spectacle will be particularly stubborn and may remain on the snake long after the remainder of the skin has been shed. In some severe cases a snake may have two or more spectacles left on the eye from previous sheddings. After shedding, the spectacles should be examined to see if they have been shed (if not, this can usually be detected by a lack of luster in the eye and the presence of a slightly ragged edge). An obstinate eye spectacle can be lubricated with glycerin or petroleum jelly over a period of two to three days, after which it can usually be gently lifted with the finger nail, taking great care not to damage the new brille beneath. On no account should a sharp metal instrument be used to do this in case the new spectacle is damaged; a completely pierced eye spectacle in snakes can quickly result in serious eye infections. Another method which can be tried is to wind a piece of adhesive tape around the finger, sticky side out, then gently dab at the brille in the hope that it will adhere to the tape. Conditions should be improved in order to diminish the possibility of troublesome sheddings recurring.

Ticks and Mites: These ectoparasites are commonly found on wild-caught snakes, and if not removed can cause serious infestations in captive stocks, resulting in anemia, shedding problems, and general debilitation. These bloodsucking parasites can also transmit some blood-borne parasitic diseases and *Aeromonas septicaemia*. Ticks attached to newly caught specimens can usually be spotted quite easily if the snake's skin is carefully examined. Up to 5mm (1/4in) in length, ticks fasten themselves with piercing mouthparts to the snake's skin, usually between the scales and often at secluded parts of the body, such as in the loose folds below the jaw, around the vent, or between the belly/subcaudal scales. Dab the tick with alcohol to relax the mouthparts before removal, otherwise there is a danger of the head being left embedded in the skin. This can result in sepsis, which is basically a synonym for infection

Mites, especially *Ophionyssus* species, are common and will become be a serious problem in the terrarium unless controlled. Unlike ticks, which stay attached to the host for some time, mites move actively from reptile to reptile and terrarium to terrarium and large numbers of them can cause stress-related debilitation, loss of appetite, and eventual death. During the daytime, mites hide away in cracks and crevices, emerging at night to suck the reptile's blood. They are about the size of a pinhead, roughly globular in shape, and normally brownish in color, but become red when bloated with blood. Mites can often be discovered moving over a reptile's body or attached to its skin between the scales if the terrarium lights are suddenly turned on at night. In severe infestations, mites may also be seen during the day. Another sign of a mite infestation is the appearance of their dustlike, silvery feces on the reptile's skin or other surfaces in the terrarium.

Infected reptiles should be moved to a clean terrarium in which is suspended a piece of dichlorvos insecticidal strip and left for a period of four days. About 1 sq cm per 10 liters ($\frac{1}{5}$ sq in per cu ft) of airspace is adequate. The strip must be placed inside a perforated container so that the reptiles cannot come into direct contact with the insecticide. All free-moving mites will be destroyed by the vapors emitted from the strip. The treatment should be repeated after 10 days to destroy any nymphs which hatch from eggs in the interim period. Vacated

Ticks can be removed easily enough with a fairly sharp lance. This of course takes a steady hand and a bit of nerve.

infected terraria should be thoroughly cleaned, scrubbed and disinfected. With very severe infestations it will be necessary to get the whole interior of the room and its contents treated.

Because dichlorvos strips may be dangerous and hard to find, many keepers now are using a solution of the synthetic pyrethmoid Permectrin (1 part to 99 parts water) sprayed on the snake and cage and then wiped off.

Endoparasites: Parasitic organisms living in the body interior are known as endoparasites. Those of most significance to the snakekeeper are various species of roundworm and tapeworm, collectively known as *helminths*. Most wild snakes play host to one or more species of intestinal worm and, in normal situations, complications are rare. However, worm infestations can become a problem in snakes suffering from stress; this could result in an abnormal increase in size or number of worms. As intestinal worms steal the snake's partly digested food, large ones or large numbers of small ones can cause anemia, general debilitation, and eventual death by starvation, possibly exacerbated by toxic wastes from the worms. Signs of severe endoparasitic infection include loss of appetite, emaciation, and the presence of worms, their eggs, or segments of tapeworms in fecal samples. Routine microscopic examination of fecal samples (which can be arranged by a veterinarian) will provide evidence of infection and its degree of seriousness. Your veterinarian will advise you on treatment for worm infections, and some of the vermicidal chemicals used for the more conventional domestic animals can be used for snakes as long as the correct dilutions and dosages are given. They may be administered via stomach tube or, if the snake is feeding readily, by first injecting the vermicide into a dead prey animal.

Protozoan Infections: The protozoan *Entamoeba invadens* and other similar organisms are sometimes responsible for dysentery-like conditions in captive snakes. Symptoms include general debilitation and watery, slimy feces. Such infections can rapidly reach epizootic proportions in a captive collection if left untreated. Effective treatments are available from your veterinarian.

Bacterial Infections: Infective salmonellosis occasionally occurs in snakes. As some salmonella species can be pathogenic to humans, the need for good hygienic practices is emphasized. Infective salmonellosis manifests itself in watery, often greenish, foul-smelling feces. Treatment with antibiotics, as advised by your vet, often proves effective. Complications of various bacterial infections include pneumonia, septicemia, and granulomatous lesions in the skin and internal organs. Early diagnosis and antibiotic treatment will often effect a complete cure.

Mouth Rot: Correctly termed necrotic or ulcerative stomatitis, this condition is unfortunately one of the most frequently encountered diseases in captive snakes and is often fatal. Stressed individuals are particularly susceptible. Nervous snakes which strike against the terrarium glass and injure the mouth and jaw are susceptible to bacterial invasion of the mucus membranes, usually by organisms of the genera *Aeromonas* or *Pseudomonas*. In severe cases the linings of the jaw margins swell to such an extent that the mouth cannot be properly closed. Adjacent bones may then become infected and death will soon follow. Death is often caused by a combination of starvation (as a snake with this condition will not feed), the toxic products of the bacterial infection, and possible infections of other parts of the body. Early diagnosis is important, and any swelling or abnormality of the mouth should be immediately investigated. Signs of infection include inflamed mucous membranes, with a gray, paste-like exudate adhering to areas around the teeth. The mouth should be held open and swabbed out (cotton swabs are ideal for this) with povidone-iodine or hydrogen peroxide, taking particular care to remove the exudate. Swabbing may be required daily over a period of several days, and it may be necessary to force-feed the snake during treatment. In advanced cases, the veterinarian may advise the surgical removal of infected tissue and bone under general anesthetic. Such treatment will usually be accompanied by a course of antibiotics.

Abscesses and Cysts: Hard or soft lumps on a snake's body under the skin may be caused by a bacterial invasion of a previous wound thought to be healed. Large abscesses or cysts can be opened surgically under anaesthetic, by a qualified veterinarian. The wound is cleaned out and swabbed with povidone-iodine and treated with an antibiotic before being sutured. Further antibiotic treatment by injection may be required during the healing process.

A good place to check for early signs of disease on any snake is around the eyes.

Reproduction and Captive Breeding

Snakes that give live birth are obviously easier for a professional breeder to deal with than those that lay eggs, which will then have to be carefully incubated.

The most gratifying and exciting element of snakekeeping is encouraging the reptiles to breed and raising the offspring to maturity. Such successes are further enhanced by captive breeding over a number of generations. However, only those specimens which are kept in optimum conditions are likely to attempt reproduction, and a detailed knowledge of a species' natural habitat and environment is a vital aspect of any serious breeding program.

Male or Female?
Determination of sex in snakes is not generally easy, as in most species there is very little sexual dimorphism. In some families the male's tail (being the distance from the posterior edge of the anal scale(s) to the tail tip) is

longer than that of the female. This difference is particularly marked in the Viperidae, many of the Crotalidae, and most members of the Boidae, in which the female has a short, sharply tapering tail, while that of the male is two to five times longer. The hemipenes, which are normally inverted into the tail base of the male, cause the tail base to be relatively thicker than that of the female. Subcaudal scale counting is a fairly reliable method of sexing many species, the scales below the tail being greater in number in males than in females. Of course, most of these methods are only reliable when you have a number of specimens for comparison. In most snake species the adult females are larger and more robust than the corresponding males, though there are exceptions. This, however, will not help if you are dealing with immature specimens.

A technique known as genital probing is a reliable modern method of sex determinarion in snakes. Probing is carried out using instruments called sexing probes, which can be obtained from specialist suppliers (ask at your local pet shop or herpetological society) and are usually available in sets containing several sizes suitable for small to large species or individuals. They may be manufactured from stainless steel or synthetic material, and basically resemble a knitting needle with a little ball at the tip. Before use, the tip should be lubricated with petroleum jelly or mineral oil. The male hemipenes normally lie inverted (inside-out) in sheaths situated in the base of the tail, just posterior to the vent. By inserting a probe into either side of the vent and pushing it gently in the direction of the tail tip, it is possible to pass it inside the inverted hemipenis for a distance several times greater than in a female. The distance varies from species to species, but normally it is possible to pass the probe for up to a distance of about seven subcaudal scales in the male, but rarely more than two or three in the female. The snake should be restrained (you may require help) and insertion into the vent carried out with great care so as not to cause injury. If one probe does not slide in easily, try a smaller one. Never force a probe home. With a little practice, the use of probes for sexing snakes becomes easy.

Breeding Cycles
Snakes are very close to and interactive with their environment, and seasonal changes are often the triggers which cause the reptiles to switch into reproductive mode. The frequency of breeding (usually once per year) and

the combination of conditions which are favorable to the breeding mode may be collectively referred to as the breeding or reproductive cycle. If kept in conditions alien to those of the natural habitat, the breeding will be confused or impaired, resulting in few breeding successes. I make no excuse for repeating the fact that it is vital to have a knowledge of a species' natural habitat if you want your snakes to breed successfully. The provision of seasonal changes in the terrarium will greatly enhance the chances of successful propagation.

Species from temperate climates will benefit from a period of hibernation followed by a progressive increase in temperature and photoperiod. Species from dry subtropical climates will still require seasonal changes in temperature and photoperiod, even if not hibernated, though a short winter period of rest at reduced temperature is always recommended. Some species may benefit from a period of estivation, others from variations in humidity. Many rain-, montane- or monsoon-forest species lay their eggs to coincide with seasonal periods of high humidity in order to facilitate satisfactory development of the embryo. By experimenting with heaters, lights, time-switches, and humidifiers, it is possible to emulate the native habitat, resulting in the successful breeding of most species.

Courtship

Outside the breeding season most snakes lead relatively solitary lives. However, the natural breeding stimuli in the appropriate season bring on the reproductive mode and make the male snakes active in seeking a mate. At the same time, this will make female snakes receptive to sexual advances. A sexually mature female emits secretions from the anal gland and leaves a scent track which is highly attractive to the male. Additionally, female snakes secrete a pheromone called vitellogen from between the scales that further stimulates male sexual activity. After encountering a female scent trail the male follows it with great determination, flickering his tongue at frequent intervals. Taking the correct direction along a scent trail is presumably decided by the progressive increase in concentration of the scent, and as a male wanders near a female along the trail he will flicker his tongue at

Facing page: Captive breeding, or *herpetoculture*, is probably the most popular aspect of today's herp hobby. Often you will be able to purchase adult snakes in pairs rather than singly.

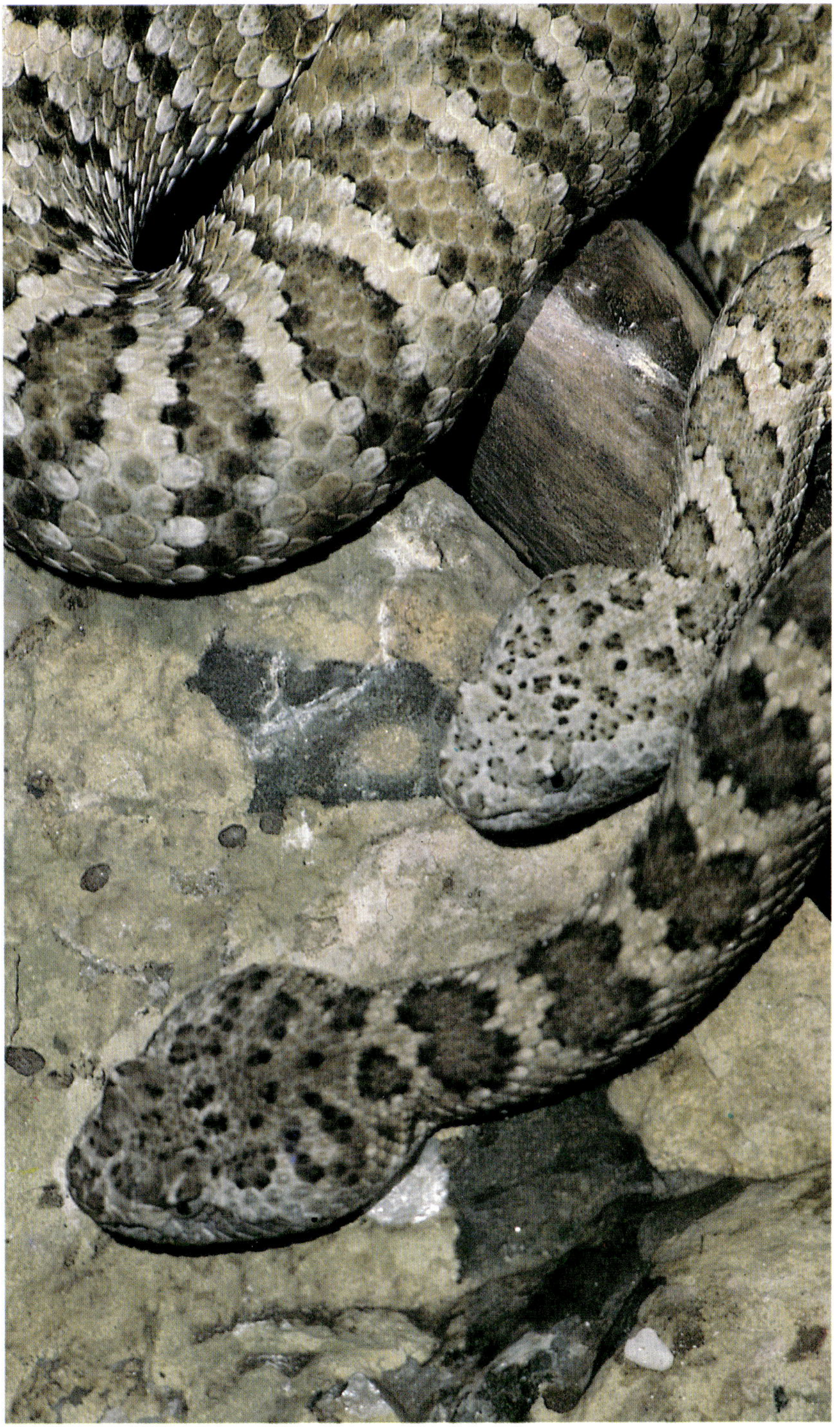

increasing intervals and the excitement in his movements becomes obvious to the onlooker.

The greatest breeding successes are attained when the sexes are kept separately for most of the year, introducing females to males at breeding time. Those specimens which are physiologically prepared through optimum climatic conditions will normally commence courtship almost immediately after introduction. In natural conditions, several males may converge on a receptive female after following her scent trail and, in some species, a combat ritual will take place in which rival males vie for the attentions of the female. A combat ritual is a form of serpentine wrestling in which males, with bodies entwined, will attempt to force the heads of their adversaries to the ground. Four, five, or even more males may participate, though two is more usual. Injuries rarely occur during these battles, and the weaker, exhausted males will eventually concede defeat and crawl off to try their luck elsewhere. The dominant male may then court and mate the female without interference from other males, but his task is far from complete.

Female snakes usually seem to ignore the rigmarole of male combat rituals and may even crawl off while the suitors are doing battle.

And, having found the female, the victorious male still needs to persuade her to mate. He approaches her with short jerky movements while continually flickering his tongue in and out. He will nudge and taste the region of her vent, from where sexually stimulating scents emerge. Eventually he will move his head slowly and jerkily along the female's body in the direction of her head and endeavor to entwine his body with hers (which is not always easy when the female is not immediately receptive). He will stimulate her body with his tongue and may even grasp her head gently in his mouth. When in the correct position, he attempts to bring his cloaca into apposition with hers. Once receptive, the female will obligingly lift her tail off the ground to give the male easier access.

Eventually, with tails and bodies entwined, the male is able to force one of his engorged hemipenes into the female's cloaca and copulation commences. Copulation in snakes can take from 30 minutes to several hours.

The Reproductive System
The major sexual organ of a male snake is a pair of hemipenes (literally: half-penises), each of which is capable of operating independently of the other though only one can be used

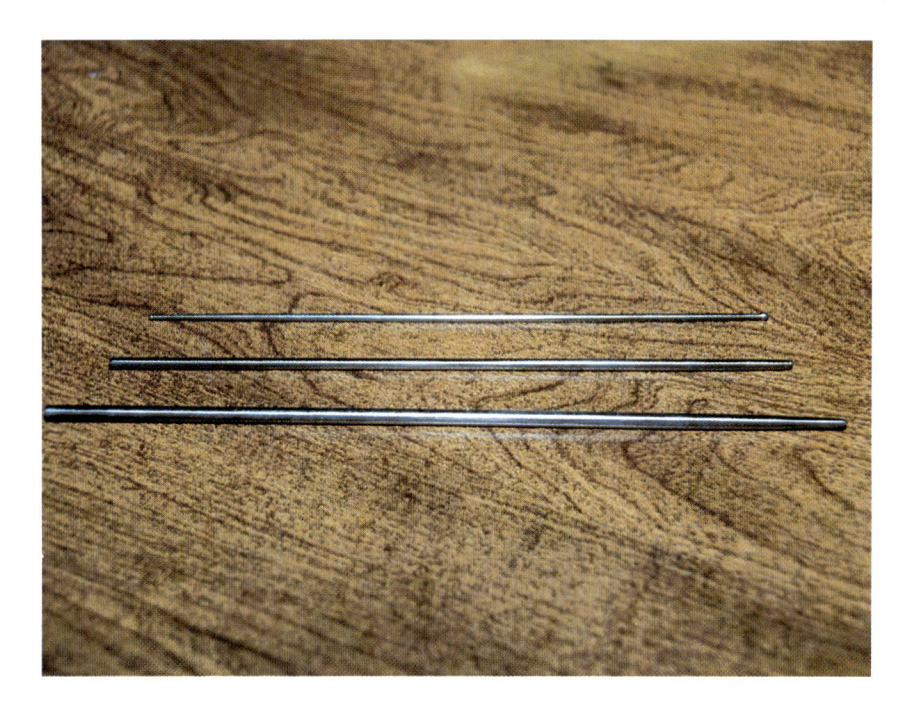

One of the most useful items for the professional snake-breeder is the sexing probe. In essence, a sexing probe is nothing more than a thin metal rod with a rounded tip. Above, three sizes are displayed. Below, a closeup of one of the ball-tips.

at a time. During non-sexual activity these organs are inverted in sheaths inside the tail base. Unlike the true penis of mammals there is no seminal duct but a simple groove along which sperm is conducted from the paired testes which are situated in the posterior part of the abdominal cavity. This groove is normally inside the sheathed hemipenis, but reaches the outer surface when the organ is everted through the vent for sexual activity. The structure of the hemipenes is usually complex and varies greatly from species to species. Many are furnished with numerous cartiliginous spines which help anchor the organ firmly in the female cloaca.

The paired female ovaries are situated in the posterior part of the body cavity. Ova are produced in the ovary and, when ripe, are then actually released into the body cavity. During ovulation in the body cavity the follicles surrounding the ova burst and the generative cells migrate to the funnel-like mouths of the paired oviducts, which open just posterior to the liver. The ova pass through the oviducts, where they are fertilized by the male spermatozoa during copulation and then stored in the uterus (which is the posterior part of the ovary). The right and left oviducts may join together first or enter the cloaca separately, depending on the species.

The Gravid Female

After fertilization, eggs are stored in the uterus where they develop over a varied period depending on the species and its method of reproduction. A female snake containing fertilized eggs is said to be **gravid**. Snakes may be **oviparous** (egglaying) or **ovoviviparous** (livebearing). In the former case the mother lays eggs containing poorly developed embryos with a large supply of nutritive yolk for the further development of the embryos within the egg but outside the maternal body. Livebearers retain the eggs in the uterus for a much longer period, and full-term embryonal development takes place within the maternal body, inside the egg which ruptures at birth or shortly after.

Some oviparous species show varying degrees of development toward ovoviviparity in that the embryos develop to a fairly advanced but not complete stage before oviposition. Thus, amongst the various oviparous species, the period of gravidity may range from 25-70 days before the eggs are deposited, while the gestation period in ovoviviparous species may take 100 days or more.

Some species are able to store viable sperm in the seminal receptacle (situated near to the head of the oviduct) for months or even years before fertilization of

One of the most exciting moments for a snake-breeder is when a pregnant mother finally lays her eggs. It is during this time that a breeder will hope to be somewhere nearby so he or she can transfer the eggs to an incubation box. The species shown is the Diadem Rat Snake, *Spalerosophis diadema*.

the ova. This explains the fact that a female snake which has been kept for a long period in captive isolation may suddenly become gravid and eventually lay a clutch of fertile eggs. Recent research also indicates that it may be an instinctive strategy of the females of some species to mate with several males so that a sort of "sperm contest" occurs in which only the best performing sperm fertilizes the eggs and thus increases the quality of the offspring. Parthenogenecity (a method of reproduction through which the ova develop without fertilization by the male gamete and produce offspring identical to the parent) has been discovered in some lizard species and in the Brahminy Blind Snake, *Ramphotyphlops braminus*. It is also suspected in a few other snake species as well.

Gravid females take on a plump appearance, particularly toward the posterior part of the body, as the eggs develop. Gravid females may fast until after egglaying or birth of the young; this is a natural phenomenon that is no cause for concern. She will drink frequently, however, so fresh water must always be available. Optimum climatic conditions are especially important at this time as a gravid female will often bask to absorb added warmth for the developing embryos. Always refrain from handling a gravid female unless it becomes absolutely necessary. Undue disturbance may result in stress that could cause many developmental problems.

Egglaying

Egglaying, sometimes referred to as oviposition, usually takes place in a location specially selected by the female as being ideal for the welfare and further development of the embryos. The major factors involved are probably temperature, humidity, and degree of concealment. In its natural wild habitat, a snake will have a great choice of egglaying sites and, as the time for oviposition approaches, will spend up to several days searching for the ideal spot. This may be under a log, in a rotten log, beneath a rock or stone, in loose soil, or among rotting vegetation. Sometimes suitable egglaying sites may be few and far between, resulting in several snakes using the same site. Communal nests containing hundreds of eggs are occasionally recorded. The interior of a termite mound is often a favored egglaying site for many snake species. Termites are able to regulate the microclimate in their nests and the resulting conditions are ideal for embryo development in the eggs of certain snake species.

Choice of egglaying sites is relatively limited in the terrarium, so it is important to try and provide something which is acceptable to the snake. Egglaying chambers seem to offer a reasonable compromise for many species. Such a chamber can easily be made using a plastic margarine or ice cream tub or a similar container compatible with the size of the snake. Holes of a diameter just large enough for the snake to gain entry are cut in one or two sides of the tub. The interior of the tub is loosely filled with slightly dampened sphagnum moss. If two or more such chambers are offered in the heated terrarium there is a good chance that a gravid female will select one in which to lay her eggs. Occasionally a snake will not be satisfied with anything you offer and eventually will scatter the eggs haphazardly about. A close watch should be kept at egglaying time so that such eggs can be quickly rescued.

Livebearing snakes are usually less selective and tend to deposit their young or hatching eggs wherever they happen to be at the time, though there is evidence to suggest that they will seek a secluded spot if possible. The young are usually brought forth in a transparent membrane which they will break open during or shortly after birth. Occasionally, usually in hot, dry conditions, the membrane will dry out too quickly and the young snakes will have difficulty in escaping. In such cases, the membrane can be carefully snipped open with a pair of surgical scissors. In most cases, young snakes are immediately active and will soon seek out hiding or basking places. Little or no maternal interest is shown by ovoviviparous snakes toward their young once they have given birth and, indeed, they may even regard them as a tasty addition to the menu. Newborn snakes are therefore best moved to separate rearing containers as soon as possible.

Incubation

In almost all cases, the best breeding results occur when the eggs of captive snakes are moved from the laying site and incubated artificially. An exception is when native snakes are being kept in an open-air walled enclosure in a suitable climatic area, but even then success is not guaranteed. In the past, the most difficult part of snake breeding has been the satisfactory incubation of the eggs. In recent years, however, a better understanding of the conditions and techniques required by developing eggs has brought the possibility of reasonable success to the serious enthusiast.

The eggs of oviparous

Notice how this mother Bullsnake, *Pituophis sayi*, attentively wraps herself around her eggs. Although many herpetologists see such an action as a measure of

aternal protection and even an attempt at mild incubation, it probably is best that
ll snake eggs be removed from the adults and incubated separately.

snakes are typically egg-shaped to elongate. They are white to cream in color and possess a soft, leathery shell designed to absorb moisture from the material in which they are incubating. Newly laid eggs often have a collapsed or dimpled appearance, but will soon fill out and become taut as moisture is absorbed. Eggs laid in the terrarium or in egglaying chambers should be carefully removed and preferably always kept the same way up. They should be arranged in rows and buried in hollows to about half their diameter in an incubation medium contained in a plastic box, thus allowing for regular inspection without undue disturbance. Occasionally eggs may be laid in a clump, and if not discovered and separated before the surface mucus dries out it will be impossible to separate them without damaging them. In such cases, the clump should be partially buried in the incubation medium so that as many as possible are visible.

Numerous artificial incubation techniques have been successful, ranging from placing the eggs on moist absorbent paper to burying them in slightly damp sand, peat or sphagnum moss, or a whole range of other materials. In recent years, a most satisfactory incubation material has been found to be vermiculite, an inert mineral which is available commercially from plant nurseries. It is sterile, retains moisture, and comes in various grades. For incubation purposes a fine grade is suitable. This can be mixed with roughly an equal amount (by weight) of water, which will be completely absorbed. If sand is used it should be washed and sterilized then partially dried out. Peat is best avoided as an incubation material since the acidity may impair normal development of the embryos.

The incubation medium ideally is placed in a plastic food box or similar, with holes made in the lid to allow air circulation (mild ventilation is also important to incubating eggs, but not cold drafts). The incubation box is then placed in an incubator which is maintained at 25-30° C. The type of incubator is immaterial as long as the correct temperature can be maintained. A simple but efficient incubator can be quite easily made at home using a ventilated wooden box with a glass door on the front, heated by a simple light bulb controlled by a thermostat. It is best to use a blue or red bulb to minimize light intensity. Alternatively a heat cable or heating pad may be used.

The eggs should develop well in a temperature range of 25-30° C, and variation in that range will do no harm. During development eggs

Incubating snake eggs is not a particularly difficult process, but it does require exact temperatures and specific substrates.

absorb moisture from the surrounding medium and increase in weight. Eggs that fail to absorb moisture are usually infertile but should not be discarded until this is absolutely certain. The contents of infertile eggs will putrefy and become discolored. As soon as this happens they should be removed and discarded. Sometimes a mold may form on the exposed surface of a developing egg. This often appears to do no harm, but to be on the safe side, the mold can be brushed gently away with a soft paint brush. The eggs should be inspected daily but handled as little as possible and left in their original positions. It may be necessary to add a small amount of water to the medium from time to time. After a few days, fertile eggs will show well-defined blood vessels through the shell and, if held up to the light (do not do this too often), the developing embryo will be seen as a dark shadow.

Incubation times vary from species to species and also depend on the temperature. Incubation at slightly lower temperatures will take longer than at higher ones. The incubation time for most oviparous species ranges from 40-70 days. The beginner to snake breeding is often excited and frustrated at the same time while waiting for the eggs to hatch, but his patience will be rewarded when he sees the first signs of hatching. As the fully developed embryo begins to absorb the bulk of the yolk, the shell may again collapse. The young snake possesses a single egg-tooth on its snout which it uses to slit open the shell from the inside. Small slits in the shell are the first signs of hatching. There may be a great variation in the amount of time from when the first slit appears to full hatching. Certain snakes may crawl out of the shell almost immediately after making the slit, but others may take two or three days to emerge from the egg. Occasionally a snake may stick its head and neck out of the shell but rapidly withdraw to the safety of the shell if disturbed. Sometimes it is a great temptation to help babies which appear to be taking a long time to hatch, but this is a natural process and must be left to the reptiles themselves. They will be absorbing the remaining contents of the yolk sac, which is attached at the lower abdomen, as well as developing the use of the lung. On rare occasions the albumen surrounding the hatchling will dry out too quickly, causing the snake to adhere to the shell. In such cases the snake can be released by gently dissolving the dried albumen with lukewarm water on a piece of cotton or paper towel.

Rearing

When the hatchlings are free-moving and detached from the eggshell, they should be removed from the incubation chamber. No attempt should be made to remove the yolk sac that may still be attached to the abdomen; this will soon shrivel up, dry, and detach itself. The tiny scar left on the belly will soon heal without treatment though it will do no harm to dab it with a mild solution of medical antiseptic. The young are best housed in small, ventilated, individual containers such as plastic lunch boxes or plastic aquaria. The containers themselves can be kept in a heated room or within a heated terrarium.

Optimum climatic requirements should be provided for the species in question and rearing containers should be simply furnished. Absorbent paper (such as kitchen towels) can be used for substrate and changed each time it becomes soiled. A small hiding box, a climbing branch, a rock, and a dish of water are sufficient, as it will be necessary to have easy access to the youngsters for inspection.

Juvenile snakes normally will not feed until after the first slough, which occurs in the first few days after birth. The frequency with which juvenile snakes feed varies greatly from species to species and even from individual to individual. Consequently some specimens grow relatively faster than others. Quite often it will be difficult to get a young snake to feed, particularly in the case of specialist feeders where every effort must be made to obtain its natural food. The important thing is to get it over the first few months.

Experimentation with different foods can begin once the snake is growing strongly. Some juvenile snakes of certain species will take food very readily, and a mistake which many beginners make is to overfeed such specimens to the point where they will become obese and malformed. Not only will these be of doubtful use for breeding but they will probably die prematurely. Juvenile snakes which readily take small mice or chicks, for example, should be fed about twice per week for the first six weeks or so, reducing feedings to once

Newborn snakes are very unpredictable, especially where feeding is concerned. Some will take food with great eagerness whereas others will have to be force-fed for the first few months.

per week thereafter. Small snakes can be given baby (pink) mice or even parts of dead adult mice (tail or limbs). Ideally, specimens should be weighed regularly and the feeding strategy adjusted so that all snakes from a particular hatching grow at a similar rate. The successful rearing of some of the more specialized feeders can constitute a challenge requiring a great deal of time and patience. The age at which snakes reach sexual maturity varies from species to species and individual to individual, and may depend on such factors as availability of food, temperature, and other climatic influences. In captivity, most species will reach breeding condition in one to three years if kept in optimum conditions.

Record Keeping

Herpetoculture is a science still in its infancy when compared with, say, aviculture. With regard to habits and reproduction in many snake species, we still have a lot to learn. It is thus important to keep records of all our experiences with snake keeping and breeding. Such records should be made available to other enthusiasts through the medium of society journals or by personal communication. It is only by combined effort and experience that progress is made in any science. Methods of record keeping will vary from individual to individual. You may prefer a system of record cards or a diary in which daily happenings are recorded. A home computer is ideal for such record keeping, especially when one requires facts at one's fingertips almost instantly.

As some species occur in very small numbers in captivity and legislation may prevent the collection of further specimens from the wild, it is obvious that an alarming amount of inbreeding is likely to occur. It is only by the keeping of records and mutual cooperation amongst snake breeders that such inbreeding can be kept to a minimum by spreading the available genes as widely as possible throughout the captive stock. With some of the rarer or endangered species it would certainly be a good idea to have a pedigree system or international studbook which could perhaps be organized by a consortium of herpetological societies. With increasing success in the numbers of snakes being bred, the breeding of color varieties has become popular. This is widely practiced among all keepers of semi-domestic animals in one form or another, so why not by snake keepers too? Many species exhibit melanism, albinism, or leucism, and color varieties now can be specially produced by captive breeding, as is especially the case with several species of *Elaphe* and *Lampropeltis*.

The most important records to be kept are those related to breeding, and such data as mating behavior, frequency of copulation, and periods of gravidity should be noted. Individual clutch records should be kept with data pertaining to the quantity, weight, and dimensions (length and breadth in mm, preferably measured with calipers) of eggs. Incubation times, temperatures, and methods should be recorded also. Hatching behavior should be noted and a record should be kept of the feeding, shedding, and growth (regular weighing and measuring) of the juveniles.

Hybridization seems to be leaving a mark in modern herpetoculture. Shown here is a cross between a Sinaloan Milk Snake, *Lampropeltis triangulum sinaloae*, and an albino Corn Snake, *Elaphe guttata guttata*.

The Human Connection

Perhaps the easiest way to "tame" a nervous snake is to handle it frequently. Although the handler may have to suffer a few bites, after a time he or she may attain the desired results.

I often am asked why I like snakes and always find it difficult to answer the question in a few words; it is perhaps easier to explain why the majority of people fear them. After all, aren't they creepy, crawly, slimy, mysterious, sinister, evil, devious, and dangerous?. The admirer of snakes will know, of course, that few of these adjectives apply to their beloved reptiles and would probably use such descriptives as interesting, fascinating, wondrous, incredible, awsome, flowing, agile, and sinuous.

People who are fond of snakes are usually also fascinated by frogs, salamanders, lizards, and turtles, all of which are studied by herpetologists. In my own case, I was attracted by creepy crawlies (as my mother called them) at a tender age. One of my earliest memories is of the fascination I experienced when observing some older boys with twigs, poking out hibernating newts from the decaying joints of a suburban wall near a local pond. That pond incidentally has long since disappeared under a concrete jungle. What really got me set on a herpetological course, however, was a few months later on vacation, seeing the capture of grass snakes by similar boys in a damp country meadow. I soon wanted a snake of my own, and after much persuasion I eventually got my father to buy me an Italian Grass Snake (probably *Natrix natrix*) from a pet shop. This of course was not allowed in the house but was consigned to an old fish tank in the garden shed. I lovingly landscaped the tank with garden soil, clods of grass, moss, and rocks, and a large soup dish as a water bath. The lid consisted of a piece of plywood with ventilation holes drilled in it and was held in place by a heavy brick. The snake was introduced to his new home and I began to collect slugs and earthworms with which to feed to him (as advised by the petshop proprietor).

Unfortunately for me, my little sister was also fascinated by the snake and a few days later, in my absence, decided to give it a cuddle without securing the lid properly after returning it to its tank. When I came back my snake was gone, never to be seen again. I was angrier than I had ever been before, but my anger turned to joy when a few days later my parents presented me with a pair of garter snakes (*Thamnophis*). I kept these for some time, feeding them on a basic diet of earthworms supplemented occasionally with small fish netted from the local waterways. They never reproduced as I was largely ignorant of their ecological requirements and maybe it was not even a true pair; there were not many accessible terrarium books around in those days (late 40's).

Since then, many books have been written about snakes and their care, and terrarium science has improved gradually over the years. Snake books are read by terrarium keepers, as well as many people who have just an interest in the mysteries of nature. Every human wants to know at least something about everything and quite often a little more about those things which frighten them, or things which they do not fully understand. So such books are read by many who rarely, if ever, see a snake that is not safely behind glass in a zoo or reptile park.

I am frequently amazed at the average person's ignorance of snakes and their habits. One would imagine especially that people who live in areas where venomous snakes are abundant would know their so-called "enemy." This is, however, rarely the case. Only recently in my local paper there was a report of a woman who found her kitten playing with a small snake and, on attempting to get the snake away, was bitten. Although the paper did not specify whether it was the cat or the snake that bit the woman, it was probably the snake as the woman was taken to hospital for treatment and released shortly thereafter, apparently none the worse for her experience. The most amazing part of the report to me, however, was the simple sentence: "The snake was not identified." In Australia, for example, where dangerously venomous snakes outnumber all others, nobody bothers to even think about getting a snake identified.

Folklore, Modern and Ancient

Throughout history, the serpentine form has been a symbol of fear to the majority of people although only about 10% of the world's 2700 or so snake species are venomous (and an even smaller percentage is dangerously venomous). Can it be that the small percentage of venomous snakes that are potential killers have given all of the others a bad reputation? This cannot be so in countries with no native venomous snakes.

But in very few countries has the creepy reputation of snakes escaped the entrepeneurs who exploit the fact to make money. Snakes appear in fiction novels, films, and on television. Live specimens are exhibited in wildlife parks and zoos and as fairground sensations and roadside attractions. Traveling snake exhibitors, whose only claim to being herpetologists is that they have been bitten several times by highly venomous snakes, make a living by sensationalizing the reptiles.

The milk snakes, *Lampropeltis triangulum*, have one of the strangest myths attached to them—it seems their common name was inspired by the belief that they sucked milk from cows. Having no lips to speak of, this must have been quite a feat.

Rattlesnakes have long been the focus of one bizarre myth or another. In the United States, these poor creatures have been utilized in thousands of quasi-religious ceremonies, often with fatal results to the worshipers. The species shown is *Crotalus durissus*.

In some countries, half-naked ladies perform sensuous dances with pythons and boas, usually in night clubs or theater clubs. There is also the story of the brave young man who makes his living mudwrestling with anacondas.

How often have you seen a snake illustrated in all its gaping, sharp-fanged, venom-dripping glory on the cover of a detective or spy novel? The text of the book may not even mention a snake, but if it does then it is likely to consist of a load of utter rubbish! Have you ever watched the reactions of friends or family as they viewed a snake in a TV movie?

Man's attitudes toward snakes are not difficult to understand, though the origins of those attitudes date back to the dawn of human life. Whether you believe in the Creation or in evolution, there were snakes around in the very early days in both cases. Indeed, though they are the most specialized members of the reptile class, snakes were around a long time before the first men appeared. Early man led a perilous existence, and while he obviously could appreciate the dangers of large predatory creatures such as lions and bears, the fact that a relatively small venomous snake could cause a dramatic and painful death would have seemed sinister and mysterious. Such mysterious properties have led certain snakes to become items of worship by many races and tribes at some time or another over the centuries.

Rock pythons were worshipped by some cultures in parts of Africa, and the killing of one was punishable by death. Other African cultures gave similar attention to other species, and various forms of snake worship were taken to the Americas during the slave-trading era. In Haiti, these forms of snake worship still manifest themselves in the practice of voodoo today. In Central America the ancient Aztecs regarded Quetzalcoatl, the plumed serpent, as the Master of Life. The Chinese imperial dragon is said to be a mixture of snake, lizard, and crocodile, while in Japan the god of thunder was portrayed as a serpent. In the village of Cuccullo in Italy snakes are paraded through the streets as a tribute to St. Dominic of Folingo, a statue of which is draped with the serpents. After the ceremony the mainly harmless (usually *Elaphe quatuorlineata*) snakes are released. In the USA many Indian tribes have worshipped snakes in one form or another, and the Pueblo Indians of the southwest consider them to be their ancestors. One of the Pueblin tribes, the Hopi, perform a famous ritual dance in honor of rattlesnakes. Snakes have also been worshipped by certain North American Christian sects since the turn of the century.

Indian cobras were (and probably still are, in certain areas) regarded by Hindu people as reincarnations of important people. These nagas (from which the generic name *Naja* is derived) were worshipped and feared by the villagers as they were capable of controlling the weather and could bring benefit or disaster alike. Anyone killing a cobra would have no male issue for 20 generations. In ancient Hindu mythology, the giant serpent Ananta had nine heads, each of which had a cobra-like hood. The well-known Buddhist Snake Temple of Penang in Malaysia contains hundreds of venomous snakes (mainly Wagler's Pit Vipers) which are fed and cared for by the monks; the temple has become a popular tourist attraction.

In Australia the Aborigines are not without their snake myths and legends. The giant rainbow serpent is associated with the creation of life. It lives in water holes which it prevents from drying up, and the snake appears in many versions of traditional aboriginal art. An aboriginal tribe local to the author's home was the Bujiebara or "people of the carpet snake." The Carpet Python, *Morelia spilotes*, was called "buji" or "booyee," and the local village of Booie derives its name from this. According to Bujiebara legend, the first Carpet Python was created in the Bunya Mountains in S.E. Queensland. The "mimburi" (source of the Carpet Pythons) was in this area, and snakes were reported to swarm over the rocky site at certain times. Women and children were not allowed to visit this mimburi in case they disturbed the snakes. An adjacent tribe, the Dungidau, believed the Bujiebara had the power to prevent the snakes entering Dungidau territory, an act which would have the effect of stopping rain in the area.

Myths and legends about snakes have been passed on from one generation of man to the next, gaining new improbabilities each time the story was told. One of the first attempts at recording such stories with the pen was made by a Syrian monk in Alexandria around 400 AD who wrote the first *Physiologus*, in which various animals and plants were given attention. It became a sort of popular biological textbook of the time and, although the original manuscript has been lost, there have been innumerable translations and retranslations into many languages. The most important of these was prepared by Aristotle and Plinius the Elder, who no doubt added their own share of improbabilities to those already recorded. References to snakes in these works show them to be capable of some amazing feats.

As snakes can move without legs, a remarkable feat by any standards, it is not surprising that such remarkable stories have arisen. In Chinese art the snake was supplied with legs and wings so that it became the dragon. However, in real life it took a long time for a dragon to mature. Hatching out as a small water snake, it took five hundred years to change into a kiao, a thousand more years to change into a lung, another five hundred years to become a k'iu-lung (horned-dragon), and yet another thousand years to become a fully fledged, winged ying-lung.

The ancient Egyptians were troubled with winged snakes that, according to Herodotus, were dangerous creatures living in trees in Arabia. They could be forced down by the burning of styrax and were then devoured by ibises (the Sacred Ibis, *Threskiornis aetheopicus*, is indeed a predator of small snakes in many parts of the world), which consequently became sacred to the Egyptians. Other Egyptian serpents, however, were divine entities. These included Sito, who encircled the earth with her immense coils while holding her tail in her mouth. In ancient China, there were apparently serpents that could entwine and kill an elephant, and youngsters of these snakes were often found in newly laid goose eggs. A Mexican serpent god, Tezcatlipoca, could swallow men whole.

In Greek legend, the demi-god Hercules had all sorts of problems with snakes. As soon as he was born he had to fight off two serpents sent to swallow him. This early herpetological experience probably came in handy later when he was commissioned to knock off the seven-headed serpent of the Lernaean Swamp. Of course, some of these early serpents may have had one head, two heads, or many heads, as in the case of the Hydra or the Medusa. In Italy, the so-called "boa" would pursue herds of cattle, cling to their udders and destroy them. Strangely, the Dhaman and the mlk snake have similar habits in India and the USA respectively. The Asp had a very bad reputation even before it bit Cleopatra; evidently it always ran about with its mouth wide open emitting steam, and it had a jewel embedded in its head. If anyone should have tried to charm it with a musical instrument it would hold one ear to the ground and block the other with the tip of its tail!

Today new myths about snakes are still arising, particularly among country folk. One of the most common of these stories is of relatively small venomous snakes hybridizing with huge pythons or boas, producing dangerously poisonous offspring of immense proportions. One of the most horrifying snakes in the USA must be the Coachwhip, which, I have been told, pursues its victim at the speed of a horse and, on catching them, binds them to a tree trunk with its coils and then proceeds to whip them to death with its tail! In Europe, French peasants believe that Grass Snakes hatch from the eggs of old roosters, and if given the opportunity the adults will suck the breasts of women. In many parts of India the harmless Dhaman (*Ptyas mucosus*) is thought to be a male cobra, and if anyone should see a Dhaman and a cobra together they will become instantly blind.

There are very many other myths, legends, and tall stories surrounding snakes, too many to include in a single volume. However, the author would like to include one more gem to whet the appetites of those wishing to further investigate this peripheral aspect of serpentology.

Taken from J. Frank Dobie (1965) is a tale that came from Georgia in 1927: "Some workmen on a power line through a swamp aroused a rattlesnake, which on striking somehow hung his fangs in the tire of an automobile. The tire must have been very thin, for the fangs penetrated far enough to puncture the inner tube. They could not be withdrawn and presently it was noticed that as the tire went down the rattlesnake was blowing up. The air pressure was presumably transmitted through the hollow fangs of the reptile. Anyway, the reptile soon became so full of air that it exploded."

Over the years, snakes have also had more practical uses. Hannibal, for example, arranged to have urns containing live venomous snakes (probably Ottoman Vipers) thrown into the Pergamanian ships, an act which brought about his victory. There are several reports of Amerindians using rattlesnakes and other venomous species as aids to the invasion of neighboring tribes. There are occasional stories in the press of venomous snakes being used to commit (usually unsuccessfully) murder or suicide. Even God used snakes to punish the children of Israel. In the Book of Numbers we find: "And the Lord sent fiery serpents among the people, and they bit the people and much people of Israel died."

Facing page: Coachwhips have a strange, if not comical, mystique about them. They earned their vernacular name from the belief that they chased people down and whipped them to death with their tails.

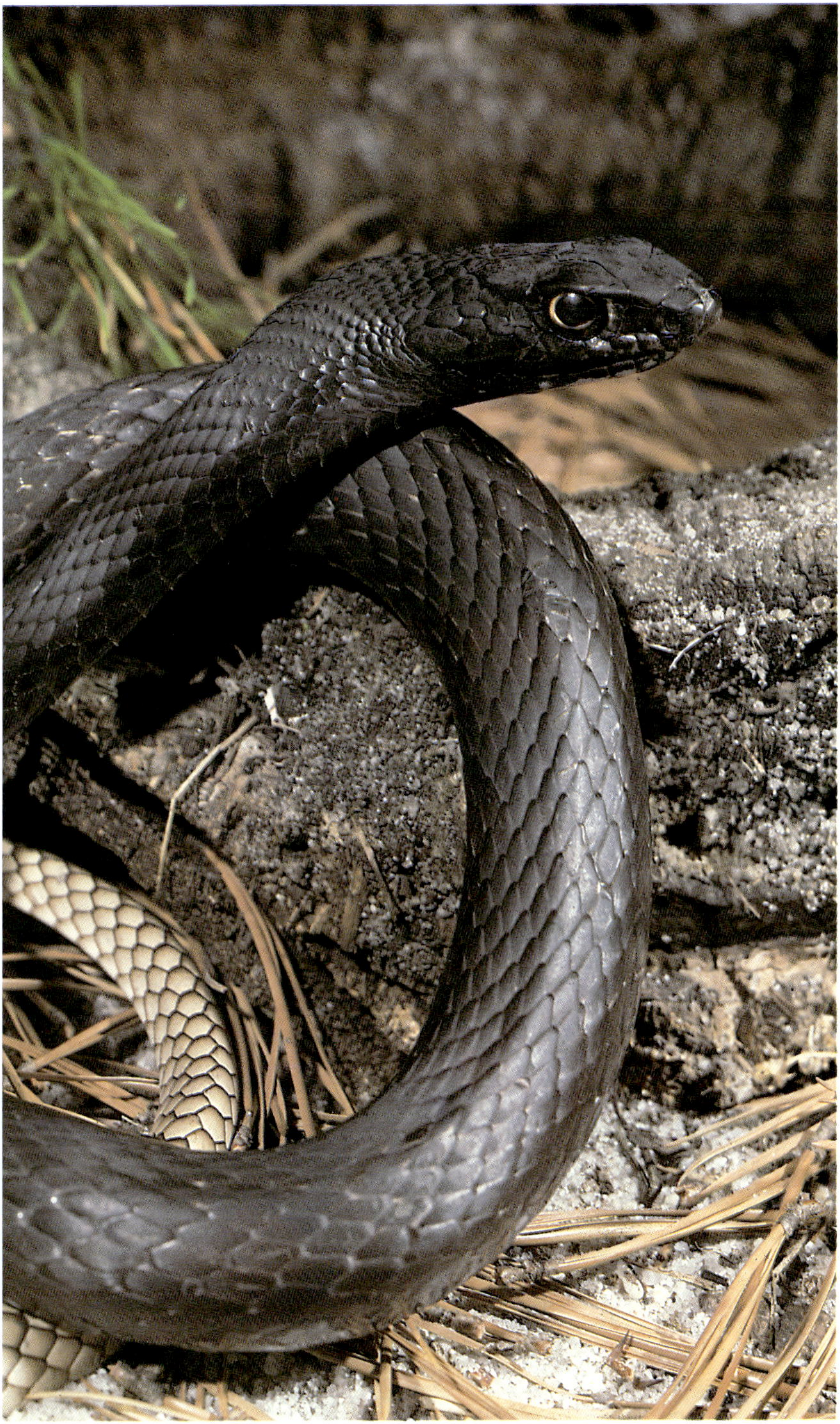

Snakes in Medicine

Over the centuries many tribes and races in all parts of the world have believed that snakes had curative attributes. To primitive man, creatures as mysterious as snakes would doubtless possess medicinal and prophylactic powers. Thus, early medicine men regularly used parts and extracts of snakes as preventions and cures for all kinds of maladies. Even today not only do medicine men in many primitive tribes still use snakes for their magical curing powers, they are also used by healers in more backward parts of civilized countries.

The ancient Greeks were among the first significant people to widely believe in the healing powers of snakes. With the ability of casting off their old skins and reappearing sleeker and apparently healthier, snakes were regarded as symbols of reincarnation and cure. Aesculapius, the Greek god of medicine, was at first portrayed as a serpent. But later, when he attained human form, he carried a staff entwined with one or two serpents. This staff is known as a "caduceus" and has today become the symbol of many organizations connected with health, hygiene, and medicine (the badge of the World Health Organization for example). The European Aesculapian Snake, *Elaphe longissima*, is so named after its apparent association with Aesculapius.

Various peoples used the healing powers of snakes in different ways. In parts of Africa, strings of snake bones may be worn by natives around their necks, wrists, or waists to strengthen parts of the body, repel evil spirits, or, paradoxically, to protect against snakebite.

Throughout history snakes have been used in Europe to treat all manner of ailments. In Europe, vipers were endowed with the most amazing medicinal powers. In one of his sermons in 1712, the Rev. W. Derham stated: "That vipers have their great uses in physick is manifest from their bearing a great share in some of our best antidotes, such as Theriaca Andromachi (a viscous potion which included parts of vipers) and others; also in the cure of elephantiasis and other like stubborn maladies, for which I shall refer to the medical writers."

Essence of vipers mixed with herbs and sometimes containing pieces of viper's flesh was widely used in Europe as a medicine. In some parts it was believed that wine fortified by

Some people have the unenviable job of milking venomous snakes for various pharmaceutical laboratories. Needless to say, this profession is highly dangerous.

preserved vipers was a cure for leprosy. In China many people made a living as snake catchers, selling their wares to apothecaries who would make all kinds of snake products to sell as cures for various ailments. Various parts and organs of snakes were, and still are, offered for the treatment of such maladies as headache, earache, toothache, convulsions, epilepsy, insanity, poor eyesight, common colds, malaria, arthritis, gout and rheumatism, and even impotence!

Various Amerindian tribes have long believed that snake fat and oil is a remedy for various complaints. European settlers in North America must have combined what they learned from the Indians with what they brought from the Old World. Rattlesnake oil was once sold throughout the USA as a home remedy for such ailments as sore throats, toothache, deafness, lumbago, and rheumatism. In Latin America snake oil is still used in poultices to cure such things as the common cold as well as various skin and joint conditions.

Modern uses of snakes by the medical profession have a more logical scientific basis than those of the past. Recent studies on the venom of various poisonous snakes have led to the production of drugs useful in the treatment of such conditions as blood disorders and arterial and heart diseases. The clotting or anticoagulant power of some

The Aesculapian Snake, *Elaphe longissima*, was named after the Greek god of medicine, Aesculapius, although the animal serves no valid medicinal purpose whatsoever.

venoms has no doubt played an important part in the selection of these venoms. As an example, the venom of the Malayan Pit Viper, *Agkistrodon (Calloselasma) rhodostoma*, has been used to produce a drug called "Arvin." The venom of this snake contains an enzyme which rapidly removes fibrinogen from the blood, thus inhibiting the formation of blood clots. Arvin is a refined form of the venom and may be used as an anticoagulant in circulatory diseases.

Venom research has led to indirect research into the captive husbandry and breeding of certain species, a fact that is useful to both the research scientist and the casual snake keeper. Research is also being conducted into the neurotoxic venoms (those possessed mainly by elapid snakes) in the hope that they may be of use in the treatment of nervous disorders.

The main reasons for keeping venomous snakes in captivity, of course (apart from showing to the public), is to have a ready supply of venom for use in the preparation of antivenom used as antidotes against snakebite.

Snake Charmers

Over the centuries, the mysterious properties of snakes have led them to be exploited by all kinds of witch doctors, magicians, and showmen as a means of making a living. The snake charmers of India are probably the best known examples in the trade. Cobras, which spectacularly spread their hoods, are favorite exhibits and are usually kept in wicker baskets. The charmer squats down in front of the basket, removes the lid, and the disturbed snake rears up in its customary threat display. If a snake should be stubborn and refuse to rear up, it will be encouraged to do so by a sharp tap on the basket or on its body. The charmer then proceeds to play musical notes on a woodwind instrument traditionally made from gourds and bamboo tubes.

Apparently mesmerized and fascinated by the music, the cobra sways from side to side. But as no snake is capable of appreciating airborne sounds, it is not the music that is doing the mesmerizing. In fact the serpent is not mesmerized at all, but behaving in the way any cobra would when threatened. It rears the front third of its body and spreads its hood in order to impress any would-be aggressor. It is in fact a warning that more unpleasant things are likely to occur if the aggressor does not go away! While playing his music the charmer moves his instrument from side to side and points it toward the snake's head. The snake follows these movements with its eyes and begins to sway in time with the moving of the pipe. That the

Facing page: It has been proven over the course of time that the music of a snake charmer's flute actually has very little to do with the snake's reaction.

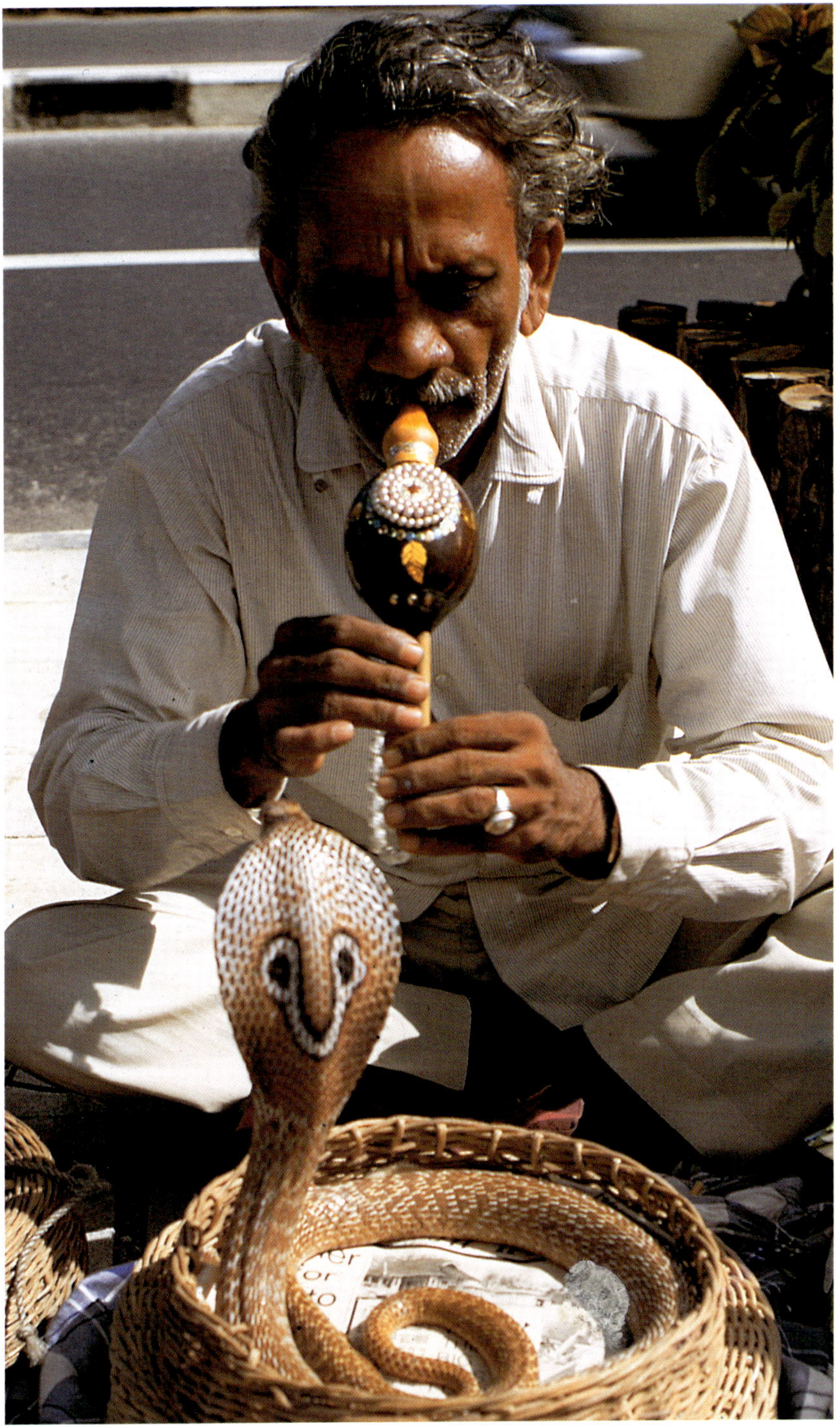

music has nothing to do with the dancing of the snake has been demonstrated many times by producing a similar reaction using a non-musical item such as a stick or an umbrella.

Lesser known but no less spectacular snake charming acts may be seen in North Africa, the Middle East, and other parts of Asia. In parts of Burma, girl charmers use King Cobras, *Ophiophagus hannah*, in an act which reaches its climax when the girl bends forward and kisses the snake on top of its head.

Snake charmers are often quite knowledgeable with regard to the care and habits of their snakes and will do their best to keep them in good condition just as a good workman looks after his tools. In areas where replacement snakes are easy to obtain, however, charmers may be more mercenary toward their charges. The act of defanging a snake or sewing its mouth shut in order to reduce the danger of a bite can only considerably reduce the reptile's life. Many authentic charmers keep their snakes in original condition, which of course increases the risk of venomous bites.

Snake Showmen and Antidote Sellers

During the latter half of the last century and the first half of the present century snake showmen (and indeed women) were a common sight at all kinds of venues. These showmen, who were mostly of European descent, could be described as modernized versions of the traditional snake charmers encountered during the period of colonization. Such showmen would take advantage of the dread that most others had for snakes, and some of them made a lot of money. Snake shows would be set up anywhere large crowds were likely to assemble, such as country fairs, amusement parks, or agricultural shows. The act usually consisted of the showman entering a snake pit which would be screened behind canvas. The pit would contain a number of snakes of various species both venomous and non-venomous. The showman would encourage the snakes to rear, rattle, hiss, or strike. He would pick up a bunch of snakes in each hand, or put the head of a snake in his mouth. He might even allow a snake to bite him and draw blood.

Snake showmen in the USA would turn up all over the country. They would frequently use rattlesnakes that had been severely mutilated by the drawing of the fangs. Many of these showmen further supplemented their income by selling antidotes for snakebite. They would claim to have been cured several times by their own antidotes, which were often a concoction of the most bizarre (but usually inexpensive) ingredients. The recipe was usually kept strictly secret and often died with its inventor.

There was never any real medical evidence that any of these early snake antidotes were indeed effective. What we do know is that in about 90% of all cases of venomous snakebite, too little venom is injected to give any serious or lasting consequences. This obviously meant that any quack antidote had a 90% chance of apparent success (providing the remedy itself did not kill the patient!). Many remedies were very expensive, but also popular and made their inventors a lot of money.

Zoos, Societies, and the Pet Trade

The latter half of the 19th century saw a boom in public zoological collections, and these aroused a wider interest in all kinds of wildlife. In major towns and cities, zoos gave urban residents the chance to view creatures that they would normally never see. Reptiles could be viewed through the glass windows of primitive terraria, most of which were little more than aquarium tanks without water. Life support systems for the inmates were primitive by today's standards, and heating arrangements often consisted of central heating of the whole reptile house with a constant high temperature (except when the system failed) day in and day out, year in and year out. Little was then known about the ecological requirements of reptiles and subsequently many of the exhibits did not survive for very long. Early zoos were thus little more than consumers of wildlife of all sorts. If a reptile died in the reptile house, it was still relatively easy to get another from the wild to replace it.

As recently as the 1950s, captive reproduction in reptiles was still the exception rather than the rule. Most captive births were from females that were already gravid when collected. Since then, however, herpetological researchers have uncovered numerous aspects of reptilian biology and ecology that have been major factors in the resulting successful captive breeding of many species through several generations. A great deal of further research with some of the more difficult species is required, however, before we can regard ourselves as thoroughly proficient in this exciting aspect of our hobby.

Major interests in herpetology have increased dramatically in recent times. Improvements in communications led to information being passed freely over greater distances, and groups of interested individuals soon got together to form clubs and societies dedicated to the study of reptiles and amphibians (for some reason, the Reptilia and the Amphibia have been lumped together in the scientific discipline of herpetology, the word being derived from the Greek "herpeton"—a creeping thing) and the furtherance of all aspects of herpetological science, ranging from field observation and ecological study to captive husbandry and breeding. Publications, symposia, and other activities have all contributed to the distribution of knowledge. Herpetology is one of those sciences in which professionals and amateurs seem to get on very well (though disputes inevitably arise on occasion) and each group agrees that much can be learned from the other. Indeed, dedicated and serious amateurs have contributed a great deal to the science. Today there is a very healthy international interest in this hitherto maligned group of vertebrates.

The rise in interest in captive herptiles produced a new generation of collectors and dealers, particularly during the 1960s when millions of tropical specimens were imported into North America and Europe. Unfortunately a huge percentage of these animals perished during transit or within two or three weeks of arrival at their destinations. This was largely due to ignorance of the collectors and transporters, and the reptiles became sick and died before reaching their final destinations.

In more recent times, the efforts of conservationists have vastly improved the fate of many species, and many countries have introduced legislation to protect wild reptiles and to improve the transporting conditions of others. Most countries now have laws relating to the capture and keeping of wild animals, including snakes, and it is beyond the scope of this book to discuss the legislation current in all those countries in which it is likely to be read. I strongly urge the reader to find out what national and local laws will affect his hobby on their home ground. It is better to be prepared rather than to risk being fined or jailed! Types of laws to be aware of are those which include aspects of keeping dangerous animals on private property, collecting from the wild, scientific experimentation on animals, zoo licensing, cruelty to animals, wildlife conservation, etc.

Snakes in Education and Science

The lower vertebrates have an important educational significance and the author has experienced many cases, particularly in primary schools, where a terrarium containing small, harmless vertebrates can help revolutionize the teaching of almost any subject. Used in a sensible manner, snakes in particular will rouse a permanent interest for small children and can be related to biology, geography, history and folklore, mathematics, chemistry, physics, art, and almost any other curricular subject.

Reptiles have played an important part in biological and evolutionary studies, and our knowledge of the latter would indeed be sparse without the possibility of studying modern living specimens. In zoology, reptiles have been popular subjects for biological studies both in the wild and in captivity. The many and varied species of snakes provide a wealth of interesting materials and it can be said that the surface of the subjects has, as yet, barely been scratched. Behavior, ecology, reproductive biology, genetics, nutrition, regeneration and healing, parthenogenesis, medicine, anatomy, morphology, physiology, and locomotion are just some of the subjects offered by snakes in man's quest for knowledge. Certain snakes breed readily in the laboratory and may be kept for various experiments. The value of primitive morphology, physiology, and behavior has recently been recognized. Metabolic studies, including the relationship of temperature and weight, have thrown light on the course of metabolic evolution and the development of homeothermy. Comparative embryological studies in particular can provide valuable information in regard to the anatomical and physiological development of the higher vertebrates, including *Homo sapiens*.

Peripheral Pastimes

People interested in snakes often have sidelines related to their main interest. A collection of snake literature is not only educational and useful, but some antiquarian snake books can be valuable items of investment. During the past few decades, many countries have produced postage stamps depicting snakes. In fact, the study of world stamp catalogues will reveal a surprising number of snake stamps. Such stamps may depict illustrations of actual snakes native to that country, or stylized snakes as symbols of organizations, etc. Collecting every snake stamp, miniature sheet, first day cover, and so on is not as easy as it may sound. Postal items depicting snakes now run into thousands and some of the older items can be quite expensive. Some of these, of course, can also be regarded as investment items.

Ornaments depicting

snakes also make a good collection. Brass and ceramic models of snakes are particularly attractive, especially those from India and China. Brass Indian cobra candlesticks are particularly collectible and these may come in many forms. Postcards depicting snakes are often issued by zoos and museums and these make another collecting possibility. Then, of course, we have paintings and prints, although not many artists tend to choose snakes as their models. But those that do produce some quite interesting and attractive collectible works.

Snake photography is a peripheral hobby which could even be profitable once you become proficient. Publishers of snake books still find it difficult to get hold of good photographs of many species, so there will usually be a market if you get some good shots. Photographing or videotaping your snakes' behavior (courting, mating, egglaying, birth, hatching, and so on) is also a very useful and interesting way of enhancing your records.

Suggested Reading

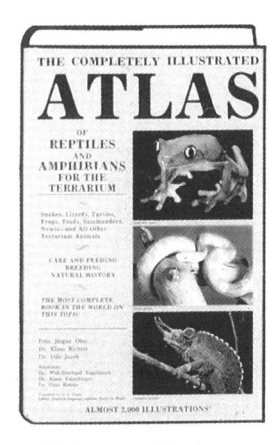

H-1102, 830 pgs, 1800+ photos

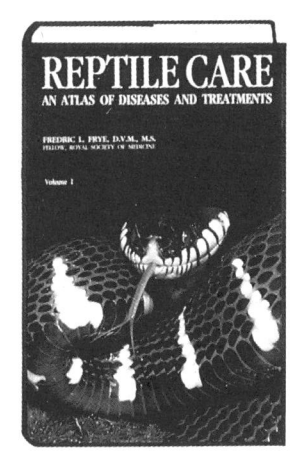

TS-165, VOL. I, 655 pgs, 1850+ photos

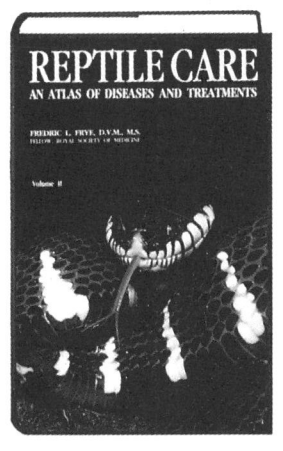

TS-165, VOL. II, 655 pgs, 1850+ photos

S-207, 230 pgs, B&W Illus.

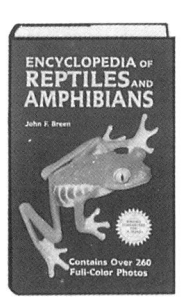

H-935, 576 pgs, 260+ photos

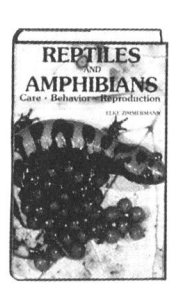

PS-876, 384 pgs, 175+ photos

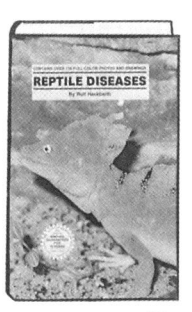

KW-197, 128 pgs, 110+ photos

TW-115, 256 pgs, 180+ photos

CO-0438, 96 pgs, 80+ photos

TU-023, 64 pgs, 50+ photos

PB-126, 64 pgs, 32+ photos

AP-925, 160 pgs, 120+ photos

TS-166, 192 pgs, 175+ photos

KW-132, 96 pgs, 40+ photos

KW-002, 96 pgs, 70+ photos

TS-154, 192 pgs, 175+ photos

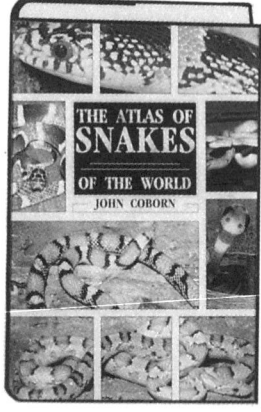

TS-128, 592 pgs, 1400+ photos

TS-125, 144 pgs, 200+ photos

TS-145, 288 pgs, 280+ photos

T-112, 64 pgs, 40+ photos

KW-127, 96 pgs, 80+ photos

PS-316, 128 pgs, 50+ photos

KW-196, 128 pgs, 100+ photos

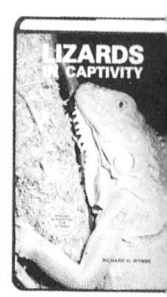

PS-769, 189 pgs, 50+ photos

SK-017, 64 pgs, 40+ photos

YF-115, 36 pgs, 20+ photos

PS-311, 96 pgs, 45+ photos

SK-015, 64 pgs, 35+ photos

INDEX

Page numbers in **boldface** refer to illustrations.

abacura abacura, Farancia, Eastern Mud Snake, **238, 239**
Abscesses, 669
Abyssinia Slug-eater, *Duberria lutrix,* 114
Acalyptophis peroni, Australian Coral Reef Snake, 453
Acanthophis antarcticus, Common Death Adder, **398, 399**
Acanthophis pyrrhus, Australian Death Adder, 399
Acrantophis dumerili, Dumeril's Boa, 50
Acrantophis madagascariensis, Madagascar Ground Boa, 50
Acrochordidae, 8
acutus, Agkistrodon [Deinagkistrodon], Chinese Copperhead, **515**, 530
adamanteus, Crotalus, Eastern Diamondback Rattlesnake, **548, 551, 558**
aegyptia, Walterinnesia, Egyptian Desert Cobra, **449, 450**
aemula, Sonora, Radiant Ground Snake, 374
aeneus, Oxybelis, American Vine Snake, 353
Aesculapean False Coral Snake, *Erythrolamprus aesculapii,* **334–335,** 336
Aesculapian Snake, *Elaphe longissima,* **701–702**
aesculapii, Erythrolamprus, Aesculapean False Coral Snake, **334–335,** 336
aestivus, Opheodrys, Rough Green Snake, 206
Afghan Awl-headed Snake, *Lytorhynchus ridgewayi,* 197
Agkistrodon bilineatus howardgloydi, Gloyd's Cantil, **528**
Agkistrodon bilineatus russeolus, Russell's Cantil, **525**
Agkistrodon bilineatus taylori, Taylor's Cantil, **513, 521**
Agkistrodon bilineatus, Tropical Cantil, **513, 514, 518, 519, 524, 526**
Agkistrodon blomhoffi brevicaudis, Short-tailed Viper, **526, 527**
Agkistrodon contortrix laticinctus, Broad-Banded Copperhead, **525**
Agkistrodon contortrix mokasen, Northern Copperhead, **522, 523, 524**
Agkistrodon contortrix, Copperhead, **521, 583**
Agkistrodon [Deinagkistrodon] acutus, Chinese Copperhead, **515,** 530
Agkistrodon [Hypnale] hypnale, Merrem's Hump-nosed Viper, **529,** 530
Agkistrodon intermedius caucasicus, Caspian Pit Viper, **528**
Agkistrodon piscivorus leucostoma, Western Cottonmouth, **516–517**

Agkistrodon piscivorus, Cottonmouth, **515, 520, 523**
Agkistrodon rhodostoma, Malaysian Moccasin, **529**
Aglyphic, 604
Ahaetulla nasuta, Long-nosed Vine Snake, 312
Ahaetulla prasina, Green Vine Snake, **15,** 312
ahaetulla, Leptophis, Parrot Snake, **195, 196**
Aipysurus apraefrontalis, New Caledonian Olive Sea Snake, 454
Aipysurus fuscus, Coastal Olive Sea Snake, 454
Aipysurus sp., Olive Sea Snake, **455**
Alameda Striped Racer, *Masticophis lateralis euryxanthus,* 201
albertisii, Liasis, D'Albert's Python, 43
albirostris, Liotyphlops, Central American Blind Snake, 34
albolabris, Trimeresurus, White-lipped Tree Viper, **569**
Alsophis vudi picticeps, Bimini Racer, 74
alterna, Lampropeltis, Gray-banded Kingsnake, **138, 139, 140, 141, 168**
alternatus, Bothrops, Urutu, 534
Amazonian Large-eyed Snake, *Thamnodynastes strigatus,* 387
Amazonian Liar, *Pseustes poecilonotus,* 221
Amazonian Short-nosed Snake, *Rhadinaea brevirostris,* 86
Amboina Yellow-lipped Snake, *Crotaphopeltis hotamboeia,* 326
American Vine Snake, *Oxybelis aeneus,* 353
ammodytes transcaucasiana, Vipera, Caucasus Sand Viper, **509**
ammodytoides, Bothrops, Patagonian Pit Viper, 534
amoenus amoenus, Carphophis, Eastern Worm Snake, 254
Amphibians, as a food item, 639
Amphiesma sp., Keeled Water Snake, 253
Amphiesma stolata, New Guinean Keeled Water Snake, **253,** 586
Amplorhinus multimaculata, Cape Many-spotted Snake, 313
Anatomy, of snakes, 591
anchietae, Python, Angola Python, 49
Andean Milk Snake, *Lampropeltis triangulum andesiana,* 172
Angola Python, *Python anchietae,* 49
angulifer, Epicrates, Cuban Boa, 670
angusticeps, Dendroaspis, Green Mamba, 412
Aniliidae, 7
Anilius scytale, False Coral Snake, **36–37**
annulata, Boulengerina, Ringed Water Cobra,

711

annulata, *Leptodeira*, Ringed Cat-eyed Snake, **346**
annulata, *Vermicella*, Bandy-bandy, **448, 449**
annulatus, *Emydocephalus*, Indonesian Turtle-headed Sea Snake, **457**
anomala, *Liophis*, Broken Ground Snake, **82**
Anomalepidae, **6**
antarcticus, *Acanthophis*, Common Death Adder, **398, 399**
"Antidote Sellers", **704**
Aparallactinae, **16**
Aparallactus capensis, Cape Centipede-eating Snake, **396**
apraefrontalis, *Aipysurus*, New Caledonian Olive Sea Snake, **454**
arenarius, *Spalerosophis*, Red-spotted Diadem Snake, **230**
argenteus, *Oxybelis*, Silver Vine Snake, **354–355**
Argentinian Bush Racer, *Oxyrhopus rhombifer*, **357, 360**
Argentinian Green Snake, *Philodryas baroni*, **85**
arietans, *Bitis*, Common Puff Adder, **477, 478, 480**
Arizona Coral Snake, *Micruroides euryxanthus*, **417**
Arizona elegans eburnata, Desert Glossy Snake, **89**
Arizona elegans philipi, Painted Desert Glossy Snake, **90**
Arizona elegans, Glossy Snake, **89**
Arizona Mountain Kingsnake, *Lampropeltis pyromelana pyromelana*, **176**
Asian Black-headed Snake, *Rhynchocalamus melanocephalus*, **252**
asper, *Bothrops*, Yellow-jawed Lancehead, **535, 537**
aspera, *Candoia*, New Guinea Viper Boa, **66, 67, 622**
Aspidelaps lubricus, Cape Coral Cobra, **400, 401**
Aspidelaps scutatus, Shield-nosed Cobra, **400, 401**
Aspidites melanocephalus, Black-headed Python, **39**
Aspidites ramsayi, Ramsay's Python, **39**
aspis, *Vipera*, European Asp, **492–493**
Atheris chloroechis, Green Bush Viper, **469**
Atheris hispidus, Laurent's Mountain Bush Viper, **468, 469**
Atheris nitschei, Great Lakes Bush Viper, **468**
Atheris squamiger, Rough-scaled Bush Viper, **24, 463, 464, 465, 466, 467, 470, 611**
Atheris superciliaris, Lowland Viper, **470**
Atlantic Central American Milk Snake, *Lampropeltis triangulum polyzona*, **185**
Atlas Mountain Montpellier Snake, *Malpolon moilensis*, **18, 350, 351**
atriceps, *Boiga*, Garden Tree Snake, **317**
atropos, *Bitis*, Mountain Adder, **479**

atrox, *Bothrops*, Fer-de-Lance, **535**
atrox, *Crotalus*, Western Diamondback Rattlesnake, **551, 552–553, 556**
aulica, *Dipsadoboa*, Royal Cat-eyed Snake, **327**
aulicus, *Lycodon*, Common Wolf Snake, **246**
aurifer, *Bothrops* [*Bothriechis*], Guatemalan Palm Viper, **533**
Aurora House Snake, *Lamprophis aurora*, **242**
aurora, *Lamprophis*, Aurora House Snake, **242**
Australian Blind Worm Snake, *Ramphotyphlops nigrescens*, **33**
Australian Copperhead, *Austrelaps superbus*, **402**
Australian Coral Reef Snake, *Acalyptophis peroni*, **453**
Australian Death Adder, *Acanthophis pyrrhus*, **399**
australis, *Pseudechis*, Mulga Snake, **442–443**
Austrelaps superbus, Australian Copperhead, **402**
austriaca, *Coronella*, Smooth Snake, **105, 106**
Azemiophinae, **22**
Azemiops feae, Fea's Viper, **462**
Baja California Rat Snake, *Elaphe* [*Bogertophis*] *rosaliae*, **125**
Ball Python, *Python regius*, **48, 612, 661, 669**
Bamboo Pit Viper, *Trimeresurus gramineus*, **566**
Banded Krait, *Bungarus fasciatus*, **403, 406**
Banded Olive Snake, *Natriciteres olivacea*, **264**
Banded Rock Rattlesnake, *Crotalus lepidus klauberi*, **556**
Banded Sand Snake, *Chilomeniscus cinctus*, **93, 94–95**
Banded sea snake, *Hydrophis* sp., **597**
Banded Water Snake, *Nerodia fasciata fasciata*, **271, 275**
Banded Wolf Snake, *Lycodon subcinctus*, **246**
Bandy-bandy, *Vermicella annulata*, **448, 449**
Barbour's Coral Snake, *Micrurus fulvius barbouri*, **420**
Bark Snake, *Hemirhagerrhis nototaenia*, **339**
baroni, *Philodryas*, Argentinian Green Snake, **85**
Beaked Sea Snake, *Enhydrina schistosa*, **458**
beetzii, *Telescopus*, Namib Tiger Snake, **386**
Behavior, of snakes, **591**
berus, *Vipera*, Northern Viper, **23, 498–499, 504, 505, 508**
bicinctus, *Hydrodynastes*, Blotched Amazon Water Snake, **339**
bicolor, *Loxocemus*, Mexican Burrowing Python, **38**
bifossatus, *Mastigodryas*, Venezuelan Tropical Racer, **204**
Big Bend Patchnose Snake, *Salvadora deserticola*, **227, 228–229**
Big-eye Snake, *Zaocys dhumnades*, **234**
Big-scaled False Coral Snake, *Pliocercus* sp., **275**

bilineatus howardgloydi, Agkistrodon, Gloyd's Cantil, **528**
bilineatus russeolus, Agkistrodon, Russell's Cantil, **525**
bilineatus taylori, Agkistrodon, Taylor's Cantil, **513, 521**
bilineatus, Agkistrodon, Tropical Cantil, **513, 514, 518, 519, 524, 526**
bilineatus, Bothrops [Bothriechis], Palm Viper, **531**
bilineatus, Elapomorphus, Two-lined Burrowing Snake, **330**
bilineatus, Masticophis, Sonoran Whipsnake, **198, 204**
Bimini Island Dwarf Boa, *Tropidophis canus curtus,* **64**
Bimini Racer, *Alsophis vudi picticeps,* **74**
Biology, of snakes, 589
Birds, as a food item, 640
biscutatus biscutatus, Trimorphodon, Common Lyre Snake, **390**
biscutatus lambda, Trimorphodon, Sonoran Lyre Snake, **390**
biscutatus vandenburghi, Trimorphodon, California Lyre Snake, **389**
biscutatus vilkinsoni, Trimorphodon, Texas Lyre Snake, **389**
biscutatus, Trimorphodon, Lyre Snake, **627**
Bitis arietans, Common Puff Adder, **477, 478, 480**
Bitis atropos, Mountain Adder, **479**
Bitis caudalis, Horned Adder, **471, 472–473, 474–475, 481, 482**
Bitis cornuta, Many-horned Adder, **484**
*Bitis gabonica rhinoceros rhinoceros,*Viper, **476**
Bitis gabonica, Gaboon Viper, **477, 479, 609, 657**
Bitis inornata, Plain Mountain Adder, **480**
Bitis nasicornis, Rhinoceros Viper, **476, 478**
Bitis peringueyi, Dwarf Puff Adder, **481, 483**
Bitis schneideri, Schneider's Puff Adder, **484**
Bitis xeropaga, Desert Mountain Adder, **483**
bitorquatus, Hoplocephalus, Pale-headed Snake, **417**
bivittata, Prosymna, Two-striped Shovel-nosed Snake, **219**
bizonus, Erythrolamprus, Slender False Coral Snake, **332–333**
Black and Yellow Rat Snake, *Spilotes pullatus,* **231**
Black Kingsnake, *Lampropeltis getula niger,* **144–145, 174**
Black Mamba, *Dendroaspis polylepis,* **412**
Black Milk Snake, *Lampropeltis triangulum gaigeae,* **182**
Black Pine Snake, *Pituophis melanoleucus lodingi,* **217**
Black Rat Snake, *Elaphe obsoleta obsoleta,* **5, 117, 128**
Black Swamp Snake, *Seminatrix pygaea,* **279**
Black-banded Cat-eyed Snake, *Leptodeira nigrofasciata,* **346**
Black-banded Coral Snake, *Micrurus nigrocinctus,* **419**
Black-barred Tree Snake, *Boiga cynodon,* **317**
Black-bellied Swamp Snake, *Hemiaspis signata,* **416**
Black-headed Python, *Aspidites melanocephalus,* **39**
Black-headed Snake, *Unechis gouldi,* **448**
Black-lined Green Snake, *Hapsidophrys lineata,* **138**
Black-lipped Cobra, *Naja melanoleuca,* **432, 434**
Black-necked Cobra, *Naja nigricollis,* **427, 429**
Black-striped Snake, *Unechis nigrostriatus,* **447**
Blackhood Snake, *Tantilla rubra cucullata,* **383**
Blackmask Racer, *Coluber constrictor latrunculus,* **100**
Blackneck Garter Snake, *Thamnophis cyrtopsis,* **301**
Blanchard's Ground Snake, *Sonora semiannulata blanchardi,* **378**
Blanchard's Milk Snake, *Lampropeltis triangulum blanchardi,* **179**
blomhoffi brevicaudis, Agkistrodon, Short-tailed Viper, **526, 527**
Blood Python, *Python curtus,* **47, 659**
Blotched Amazon Water Snake, *Hydrodynastes bicinctus,* **339**
Blotched Kingsnake, *Lampropeltis getula goini,* **142**
Blue-banded Sea Snake, *Hydrophis cyanocinctus,* **458, 459**
Bluestripe Garter Snake, *Thamnophis sauritus nitae,* **284, 291, 301**
Bluestripe Ribbon Snake, *Thamnophis sirtalis similis,* **294**
Blunt-headed Tree Snake, *Imantodes cenchoa,* **342**
Boa constrictor, Boa constrictor, **51, 52, 595, 661**
Boa constrictor, *Boa constrictor,* **51, 52, 595, 661**
boa, Liasis, Ringed Python, **44**
Boaedon fuliginosus, Common House Snake, **236, 237**
Bocourt's Lance-headed Snake, *Bothrops ophryomegas,* **540**
boeleni, Python, Boelen's Python, **9, 46**
Boelen's Python, *Python boeleni,* **9, 46**
Boidae, 8
Boiga atriceps, Garden Tree Snake, **317**
Boiga cyanea, Green Cat-eyed Snake, **313**
Boiga cynodon, Black-barred Tree Snake, **317**
Boiga dendrophila, Mangrove Snake, **314, 316, 653**
Boiga fusca, Tan Tree Snake, **318**
Boiga irregularis, Brown Tree Snake, **314, 315**
Boiga multomaculata, Many Spotted Tree Snake, **318**
Boiga ocellata, Ocellated Tree Snake, **17, 315**

713

Boiginae, 15
Boinae, 10
Bolivian Green Snake, *Philodryas psammophideus*, 85
Bolyeriinae, 10
Boomslang, *Dispholidus typus*, 328, 329, 330
bornmulleri, Vipera, Bornmuller's Viper, 506
Bornmuller's Viper, *Vipera bornmulleri,* 506
Bothrops alternatus, Urutu, 534
Bothrops ammodytoides, Patagonian Pit Viper, 534
Bothrops asper, Yellow-jawed Lancehead, 535, 537
Bothrops atrox, Fer-de-Lance, 535
Bothrops [Bothriechis] aurifer, Guatemalan Palm Viper, 533
Bothrops [Bothriechis] bilineatus, Palm Viper, 531
Bothrops brazili, Brasil's Pit Viper, 538
Bothrops jararaca, Jararaca, 539, 546
Bothrops lateralis, Yellow-lipped Palm Viper, 543, 545
Bothrops moojeni, Caissaca, 546
Bothrops nasutus, Hog-nosed Pit Viper, 542
Bothrops neuwiedi, Neuwied's Pit Viper, 544
Bothrops nummifer, Jumping Viper, 538, 547
Bothrops ophryomegas, Bocourt's Lance-headed Snake, 540
Bothrops rowleyi, Rowley's Pit Viper, 542
Bothrops schlegeli, Eyelash Viper, 532, 533, 536, 537, 541, 543, 544, 545, 547, 645, 656
bottae, Charina, Rubber Boa, 68–69, 70
Boulengerina annulata, Ringed Water Cobra, 403
bovalli, Rhinobothryum, Costa Rican Tree Snake, 373
brachystoma, Thamnophis, Shorthead Garter Snake, 287
Branches, 631
Brasil's Pit Viper, *Bothrops brazili,* 538
brazili, Bothrops, Brasil's Pit Viper, 538
Brazilian Bush Racer, *Oxyrhopus trigeminus,* 360
Brazilian Liar, *Pseustes sulphureus,* 221
Breeding cycles, 671
brevicaudus, Unechis, Short-tailed Snake, 447
brevirostris, Oxybelis, Short-nosed Vine Snake, 353
brevirostris, Rhadinaea, Amazonian Short-nosed Snake, 86
Broad-Banded Copperhead, *Agkistrodon contortrix laticinctus,* 525
Broad-banded Water Snake, *Nerodia fasciata confluens,* 269, 271
Broken Ground Snake, *Liophis anomala,* 82
Brown Sand Boa, *Eryx johni,* 71
Brown Snake, *Storeria dekayi,* 280, 281
Brown Tree Snake, *Boiga irregularis,* 314, 315
Brown-spotted Night Viper, *Tomodon dorsatus,* 388
browni lucidus, Phyllorhynchus, Maricopa

Leafnose Snake, 209
browni, Phyllorhynchus, Saddled Leafnose Snake, 209
Bullsnake, *Pituophis sayi,* 215, 218, 680–681
Bungarus caeruleus, Indian Krait, 405, 406
Bungarus candidus, Malayan Krait, 407, 408
Bungarus fasciatus, Banded Krait, 403, 406
Bungarus flaviceps, Red-headed Krait, 407
Bungarus multicinctus, Many-banded Krait, 404, 405
Bushmaster, *Lachesis muta,* 560
butleri, Thamnophis, Butler's Garter Snake, 284
Butler's Garter Snake, *Thamnophis butleri,* 284
Cacophis squamulosus, Golden-crowned Snake, 408
caeruleus, Bungarus, Indian Krait, 405, 406
Caissaca, *Bothrops moojeni,* 546
Calabar Burrowing Python, *Calabaria reinhardti,* 40, 638
Calabaria reinhardti, Calabar Burrowing Python, 40, 638
Calamarinae, 11
California Kingsnake, *Lampropeltis getula californiae,* 142, 143, 146, 163, 170
California Lyre Snake, *Trimorphodon biscutatus vandenburghi,* 389
California Striped Racer, *Masticophis lateralis lateralis,* 203
calligaster calligaster, Lampropeltis, Prairie Kingsnake, 166
calligaster occipitolineata, Lampropeltis, South Florida Mole Kingsnake, 167
calligaster rhombomaculata, Lampropeltis, Mole Snake, 167
calligaster, Dendrelaphis, Northern Tree Snake, 108
Calliophis macclellandi, MacClelland's Oriental Coral Snake, 409
Calliophis sauteri, Sauter's Oriental Coral Snake, 409
cana, Pseudaspis, Molslang, 220
candidus, Bungarus, Malayan Krait, 407, 408
Candoia aspera, New Guinea Viper Boa, 66, 67, 622
Candoia carinata paulsoni, Solomon Islands Ground Boa, 66
Canebrake Rattlesnake, *Crotalus horridus atricaudatus,* 557
caninus, Corallus, Emerald Tree Boa, 9, 53, 54, 619
canum, Gyalopion, Western Hooknose Snake, 338
canus curtus, Tropidophis, Bimini Island Dwarf Boa, 64
Cape Centipede-eating Snake, *Aparallactus capensis,* 396
Cape Cobra, *Naja nivea,* 433
Cape Coral Cobra, *Aspidelaps lubricus,* 400, 401
Cape File Snake, *Mehelya capensis,* 248
Cape Many-spotted Snake, *Amplorhinus*

multimaculata, 313
Cape Twig Snake, *Thelotornis kirtlandi capensis,* 387
Cape Wolf Snake, *Lycophidion capense,* 247
capense, Lycophidion, Cape Wolf Snake, 247
capensis, Aparallactus, Cape Centipede-eating Snake, 396
capensis, Mehelya, Cape File Snake, 248
carinata paulsoni, Candoia, Solomon Islands Ground Boa, 66
carinata, Elaphe, Stinking Goddess, 134, 136–137
carinata, Pythonodipsas, Keeled Snake, 370
carinatus, Echis, Saw-scaled Viper, 487, 488, 489
carinatus, Pareas, Keeled Slug Snake, 394
carinicauda, Helicops, Common Keelback Snake, 261
Carolina Pygmy Rattlesnake, *Sistrurus miliarius miliarius,* 562
Carpet Python, *Morelia spilotes variegata,* 45
Carpet Viper, *Echis coloratus,* 486, 487, 488
Carphophis amoenus amoenus, Eastern Worm Snake, 254
Caspian Pit Viper, *Agkistrodon intermedius caucasicus,* 528
Cat-eyed Snake, *Leptodeira septentrionalis,* 347
catenatus catenatus, Sistrurus, Eastern Massasauga, 562
catenatus, Sistrurus, Massasauga, 561
catenifer affinis, Pituophis, Sonoran Gopher Snake, 215, 631
catenifer catenifer, Pituophis, Pacific Gopher Snake, 211
catesbyi, Dipsas, Catesby's Snail-sucker, 392
Catesby's Snail-sucker, *Dipsas catesbyi,* 392
Caucasus Sand Viper, *Vipera ammodytes transcaucasiana,* 509
Caucasus Viper, *Vipera kaznakovi,* 505, 506
caudalis, Bitis, Common Puff Adder, 471
caudalis, Bitis, Horned Adder, 472–473, 474–475, 481, 482
Causinae, 23
Causus defilippi, DeFilipp's Adder, 511
Causus maculatus, Spotted Adder, 511
Causus rhombeatus, Rhombic Night Adder, 510, 512
Cemophora coccinea coccinea, Florida Scarlet Snake, 90
Cemophora coccinea copei, Northern Scarlet Snake, 91
Cemophora coccinea, Scarlet Snake, 13, 91, 92
cenchoa, Imantodes, Blunt-headed Tree Snake, 342
cenchria, Epicrates, Rainbow Boa, 55
Central American Blind Snake, *Liotyphlops albirostris,* 34
Central American Snail-eater, *Sibon nebulata,* 393
Central Asian Cobra, *Naja oxiana,* 429

Central Baja Rosy Boa, *Lichanura trivirgata "myriolepis",* 62
Central Plains Milk Snake, *Lampropeltis triangulum gentilis,* 182
Cerastes cerastes, Horned Sand Viper, 486
cerastes laterorepens, Crotalus, Colorado Desert Sidewinder, 558
Cerastes vipera, Common Sand Viper, 485
cerastes, Cerastes, Horned Sand Viper, 486
Cerberus rhynchops, Dog-faced Water Snake, 309
Charina bottae, Rubber Boa, 68–69, 70
Checkered Garter Snake, *Thamnophis marcianus,* 289, 295
Checkered Keelback, *Xenochrophis piscator,* 308
Chiapan Boa, *Ungaliophis continentalis,* 65
Chihuahua Mountain Kingsnake, *Lampropeltis pyromelana knoblochi,* 162, 175
Chilomeniscus cinctus, Banded Sand Snake, 93, 94–95
Chilorhinophis gerardi, Tanganyikan Two-headed Snake, 397
chinensis, Enhydris, Chinese Water Snake, 309
Chinese Copperhead, *Agkistrodon [Deinagkistrodon] acutus,* 515, 530
Chinese Tiger Snake, *Rhabdophis tigrina,* 278
Chinese Water Snake, *Enhydris chinensis,* 309
Chionactis occipitalis klauberi, Tucson Shovelnose Snake, 97
Chionactis occipitalis, Western Shovelnose Snake, 96
Chionactis palarostris, Sonoran Shovelnose Snake, 96, 97
chloroechis, Atheris, Green Bush Viper, 469
Chondropython viridis, Green Tree Python, 41, 42
Chrysopelea ornata, Golden Flying Snake, 319, 320, 321
cinctus, Chilomeniscus, Banded Sand Snake, 93, 94–95
Circulatory system, 610
clarkii, Nerodia, Salt Marsh Snake, 270, 273
Classification table, 587
Clelia clelia clelia, East Guatemalan Mussurana, 323
clelia clelia, Clelia, East Guatemalan Mussurana, 323
Clelia clelia, Guatemalan Mussurana, 323
Clelia occipitolutea, Yellow-headed Mussurana, 322
Clelia rustica, Smooth Brown Mussurana, 19, 322
clelia, Clelia, Guatemalan Mussurana, 323
Coachwhip, *Masticophis flagellum,* 199, 697
Coastal Mountain Kingsnake, *Lampropeltis zonata multifasciata,* 190, 192
Coastal Olive Sea Snake, *Aipysurus fuscus,* 454
Coastal Plains Milk Snake, *Lampropeltis triangulum "temporalis",* 156

715

coccinea coccinea, Cemophora, Florida Scarlet Snake, **90**
coccinea copei, Cemophora, Northern Scarlet Snake, **91**
coccinea, Cemophora, Scarlet Snake, **13**, **91**, **92**
Collared Dwarf Snake, *Eirenis collaris,* **115**
collaris, Eirenis, Collared Dwarf Snake, **115**
Colorado Desert Sidewinder, *Crotalus cerastes laterorepens,* **558**
coloratus, Echis, Carpet Viper, **486**, **487**, **488**
Coluber constrictor constrictor, Northern Black Racer, **103**, **104**
Coluber constrictor flaviventris, Eastern Yellowbelly Racer, **99**
Coluber constrictor latrunculus, Blackmask Racer, **100**
Coluber constrictor mormon, Western Yellowbelly Racer, **103**
Coluber constrictor priapus, Southern Black Racer, **104**
Coluber hippocrepis hippocrepis, Horseshoe Racer, **98**, **651**
Coluber jugularis, Large Whip Snake, **102**
Coluber karelini mintonorum, Minton's Spotted Mountain Racer, **102**
Coluber najadum dahli, Dahl's Whipsnake, **101**
Coluber ravergieri, Mountain Racer, **98**, **100**, **101**
Coluber sp., racer, **599**
Colubridae, **10**
colubrina, Laticauda, Yellow-lipped Sea Krait, **451**, **453**
Colubrinae, **12**
colubrinus loveridgei, Eryx, Sri Lankan Sand Boa, **72**
colubrinus, Madagascarophis, Madagascan Blunt-nosed Snake, **349**, **350**
Common Black-banded Coral Snake, *Micrurus nigrocinctus nigrocinctus,* **425**
Common Cobra, *Naja naja,* **426**, **432**
Common Coral Snake, *Micrurus corallinus corallinus,* **421**
Common Death Adder, *Acanthophis antarcticus,* **398**, **399**
Common Egg-eating Snake, *Dasypeltis scabra,* **235**, **641**
Common Garter Snake, *Thamnophis sirtalis,* **286**, **293**, **299**
Common Green Bush Snake, *Philothamnus hoplogaster,* **207**
Common House Snake, *Boaedon fuliginosus,* **236**, **237**
Common Keelback Snake, *Helicops caricauda,* **261**
Common Lined Snake, *Tropiclonion lineatum lineatum,* **304–305**
Common Lyre Snake, *Trimorphodon biscutatus biscutatus,* **390**
Common Puff Adder, *Bitis arietans,* **477**, **478**, **480**
Common Sand Viper, *Cerastes vipera,* **485**

Common Tree Snake, *Dendrelaphis punctulatus,* **107**
Common Wolf Snake, *Lycodon aulicus,* **246**
Conant's Milk Snake, *Lampropeltis triangulum conanti,* **161**, **181**
Coniophanes imperialis, Royal Black-striped Snake, **324–325**
Coniophanes piceivittis, Double Black-striped Snake, **326**
Conopsis lineatus, Queretaro Ground Snake, **75**
Constriction, **606**
constrictor constrictor, Coluber, Northern Black Racer, **103**, **104**
constrictor flaviventris, Coluber, Eastern Yellowbelly Racer, **99**
constrictor latrunculus, Coluber, Blackmask Racer, **100**
constrictor mormon, Coluber, Western Yellowbelly Racer, **103**
constrictor priapus, Coluber, Southern Black Racer, **104**
constrictor, Boa, Boa constrictor, **51**, **52**, **595**, **661**
Contia tenuis, Sharp-tailed Snake, **105**
continentalis, Ungaliophis, Chiapan Boa, **65**
contortrix laticinctus, Agkistrodon, Broad-Banded Copperhead, **525**
contortrix mokasen, Agkistrodon, Northern Copperhead, **522**, **523**, **524**
contortrix, Agkistrodon, Copperhead, **521**, **583**
Copperhead, *Agkistrodon contortrix,* **521**, **583**
Coprodeum, **609**
corais corais, Drymarchon, Yellow-tailed Indigo Snake, **110**
corais couperi, Drymarchon, Eastern Indigo Snake, **110**, **111**
corais erebennus, Drymarchon, Texas Indigo Snake, **109**
corais melanurus, Drymarchon, Gray Indigo Snake, **109**, **111**
Coral Cobra, *Micrurus frontalis frontalis,* **421**
Coral Snake, *Micrurus fulvius,* **418**, **420**, **422–423**, **425**
Coralbelly Ringneck Snake, *Diadophis punctatus pulchellus,* **257**
corallinus corallinus, Micrurus, Common Coral Snake, **421**
Corallus caninus, Emerald Tree Boa, **9**, **53**, **54**, **619**
Corallus enydris, Garden Boa, **53**, **54**
Corn Snake, *Elaphe guttata guttata,* **117**, **119**, **129**, **130**, **131**, **132–133**, **581**, **633**
cornuta, Bitis, Many-horned Adder, **484**
Coronella austriaca, Smooth Snake, **105**, **106**
Coronella girondica, Southern Smooth Snake, **106**
coronoides, Drysdalia, White-lipped Snake, **413**
Costa Rican Tree Snake, *Rhinobothryum bovalli,* **373**
Cottonmouth, *Agkistrodon piscivorus,* **515**,

716

520, 523
Courtship, 672
Crickets, as a food item, 635
Croatlus atrox, Western Diamondback Rattlesnake, 556
crossi, Mehelya, Fringed File Snake, 248
Crotalidae, 23
Crotalus adamanteus, Eastern Diamondback Rattlesnake, 548, 551, 558
Crotalus atrox, Western Diamondback Rattlesnake, 551, 552–553
Crotalus cerastes laterorepens, Colorado Desert Sidewinder, 558
Crotalus durissus, Neotropical Rattlesnake, 692
Crotalus horridus atricaudatus, Canebrake Rattlesnake, 557
Crotalus horridus horridus, Timber Rattlesnake, 5, 559, 699
Crotalus lepidus klauberi, Banded Rock Rattlesnake, 556
Crotalus mitchelli pyrrhus, Southwestern Speckled Rattlesnake, 550
Crotalus mitchelli, Speckled Rattlesnake, 673
Crotalus molossus, Northern Black-tailed Rattlesnake, 550
Crotalus ruber, Red Diamond Rattlesnake, 554–555
Crotalus viridis oreganus, Northern Pacific Rattlesnake, 549, 557
Crotaphopeltis hotamboeia, Amboina Yellow-lipped Snake, 326
Cryptophis nigrescens, Eastern Small-eyed Snake, 410
Cuban Black-tailed Dwarf Boa, *Tropidophis melanurus melanurus,* 64
Cuban Boa, *Epicrates angulifer,* 670
cucullatus, Macroprotodon, False Smooth Snake, 348
Curl Snake, *Suta suta,* 446
curtus brongersmai, Python, Malayan Blood Python, 48
curtus, Python, Blood Python, 47, 659
cyanea, Boiga, Green Cat-eyed Snake, 313
cyanocinctus, Hydrophis, Blue-banded Sea Snake, 458, 459
cyclopion, Nerodia, Mississippi Green Water Snake, 270
cynodon, Boiga, Black-barred Tree Snake, 317
cyrtopsis, Thamnophis, Blackneck Garter Snake, 301
Cysts, 669
Dahl's Whipsnake, *Coluber najadum dahli,* 101
D'Albert's Python, *Liasis albertisii,* 43
Dasypeltinae, 12
Dasypeltis inornata, South African Egg-eating Snake, 236, 603
Dasypeltis scabra, Common Egg-eating Snake, 235, 641
decurtus perkinsi, Phyllorhynchus, Western Leafnose Snake, 208
decurtatus, Phyllorhynchus, Spotted Leafnose Snake, 210
defilippi, Causus, DeFilipp's Adder, 511
DeFilipp's Adder, *Causus defilippi,* 511
dekayi victa, Storeria, Florida Brown Snake, 282–283
dekayi, Storeria, Brown Snake, 280, 281
Demansia psammophis, Yellow-faced Whip Snake, 411
Dendrelaphis calligaster, Northern Tree Snake, 108
Dendrelaphis pictus, Indonesian Bronzeback, 107
Dendrelaphis punctulatus, Common Tree Snake, 107
Dendroaspis angusticeps, Green Mamba, 412
Dendroaspis jamesoni, Jameson's Mamba, 21, 411
Dendroaspis polylepis, Black Mamba, 412
dendrophila, Boiga, Mangrove Snake, 314, 316, 653
Denisonia punctata, Little Spotted Snake, 413
deppei deppei, Pituophis, Mexican Pine Snake, 216
deppei jani, Pituophis, Jan's Pine Snake, 216
Desert Glossy Snake, *Arizona elegans eburnata,* 89
Desert Hooknose Snake, *Gyalopion quadrangularis,* 338
Desert Kingsnake, *Lampropeltis getula splendida,* 169
Desert Mountain Adder, *Bitis xeropaga,* 483
Desert Rosy Boa, *Lichanura trivirgata gracia,* 61
deserticola, Salvadora, Big Bend Patchnose Snake, 227, 228–229
dhara, Telescopus, Large-eyed Snake, 384
dhumnades, Zaocys, Big-eye Snake, 234
Diadem Rat Snake, *Spalerosophis diadema,* 230, 677
diadema, Furina, Red-naped Diadem Snake, 414
diadema, Spalerosophis, Diadem Rat Snake, 230, 677
Diadophis punctatus amabilis, Pacific Ringneck Snake, 259
Diadophis punctatus arnyi, Prairie Ringneck Snake, 258
Diadophis punctatus pulchellus, Coralbelly Ringneck Snake, 257
Diadophis punctatus punctatus, Southern Ringneck Snake, 257
Diadophis punctatus regalis, Regal Ringneck Snake, 256
Diadophis punctatus, Ringneck Snake, 255, 256
Diamondback Water Snake, *Nerodia rhombifer rhombifer,* 274
Diet, range of, 605
Digestion, 602, 608
Dinodon rufozonatum, Red-banded Snake, 237
diplotrophis, Leptophis, Sunrise Parrot Snake,

717

Dipsadinae, 16
Dipsadoboa aulica, Royal Cat-eyed Snake, 327
Dipsadoboa pulverulentus, Dusky Tropical Cat-eyed Snake, 327
Dipsas catesbyi, Catesby's Snail-sucker, 392
Dipsas sp., Snail-sucker, 392
Dispholidus typus, Boomslang, 328, 329, 330
Disteira [Astrotia] stokesi, Stoke's Sea Snake, 455, 456
Dixon's Milk Snake, *Lampropeltis triangulum dixoni,* 181
Dog-faced Water Snake, *Cerberus rhynchops,* 309
dorbignyi, Lystrophis, Dorbigny's Hognose Snake, 83, 84
Dorbigny's Hognose Snake, *Lystrophis dorbignyi,* 83, 84
dorsalis, Uromacer, Haitian Longtail Snake, 87
dorsatus, Tomodon, Brown-spotted Night Viper, 388
Double Black-striped Snake, *Coniophanes piceivittis,* 326
Drymarchon corais corais, Yellow-tailed Indigo Snake, 110
Drymarchon corais couperi, Eastern Indigo Snake, 110, 111
Drymarchon corais erebennus, Texas Indigo Snake, 109
Drymarchon corais melanurus, Gray Indigo Snake, 109, 111
Drymobius margaritiferus, Mexican Speckled Snake, 112–113, 114
Drysdalia coronoides, White-lipped Snake, 413
Duberria lutrix, Abyssinia Slug-eater, 114
dulcis dissectus, Leptotyphlops, New Mexico Blind Snake, 35
dulcis, Leptotyphlops, Texas Blind Snake, 7
dumerili, Acrantophis, Dumeril's Boa, 50
dumerili, Micrurus, Dumeril's Coral Snake, 419
Dumeril's Boa, *Acrantophis dumerili,* 50
Dumeril's Coral Snake, *Micrurus dumerili,* 419
Durango Mountain Kingsnake, *Lampropeltis mexicana "greeri",* 147, 150, 174
Durango Mountina Kingsnake, *Lampropeltis mexicana "greeri",* 151
durissus, Crotalus, Neotropical Rattlesnake, 692
Dusky Tropical Cat-eyed Snake, *Dipsadoboa pulverulentus,* 327
Dwarf Puff Adder, *Bitis peringueyi,* 481, 483
Earthworms, as a food item, 636
East African Blind Worm Snake, *Typhlops schlegeli,* 33
East Guatemalan Mussurana, *Clelia clelia clelia,* 323
Eastern Brown Snake, *Pseudonaja textilis,* 446
Eastern Chain Kingsnake, *Lampropeltis getula getula,* 172
Eastern Coachwhip, *Masticophis flagellum flagellum,* 200

Eastern Diamondback Rattlesnake, *Crotalus adamanteus,* 548, 551, 558
Eastern Garter Snake, *Thamnophis sirtalis sirtalis,* 288, 290, 298
Eastern Hognose Snake, *Heterodon platirhinos,* 12, 78–79, 80, 583, 634
Eastern Indigo Snake, *Drymarchon corais couperi,* 110, 111
Eastern Massasauga, *Sistrurus catenatus catenatus,* 562
Eastern Milk Snake, *Lampropeltis triangulum triangulum,* 152, 188, 658
Eastern Mud Snake, *Farancia abacura abacura,* 238, 239
Eastern Small-eyed Snake, *Cryptophis nigrescens,* 410
Eastern Worm Snake, *Carphophis amoenus amoenus,* 254
Eastern Yellowbelly Racer, *Coluber constrictor flaviventris,* 99
Echis carinatus, Saw-scaled Viper, 487, 488, 489
Echis coloratus, Carpet Viper, 486, 487, 488
Ecuadorian Milk Snake, *Lampropeltis triangulum micropholis,* 183
Education, snakes in, 706
Egglaying, 677, 678
Eggs, of python, 683
Egyptian Cobra, *Naja haje,* 435
Egyptian Desert Cobra, *Walterinnesia aegyptia,* 449, 450
Eirenis collaris, Collared Dwarf Snake, 115
Eirenis rothi, Roth's Dwarf Snake, 115
Elachistodontinae, 16
Elaphe [Bogertophis] rosaliae, Baja California Rat Snake, 125
Elaphe [Bogertophis] subocularis, Trans-Pecos Rat Snake, 116, 122, 685
Elaphe carinata, Stinking Goddess, 134, 136–137
Elaphe guttata emoryi, Great Plains Rat Snake, 120
Elaphe guttata guttata, Corn Snake, 117, 119, 129, 130, 131, 132–133, 581, 633
Elaphe helenae, Trinket Snake, 134
Elaphe longissima, Aesculapian Snake, 701–702
Elaphe mandarina, Mandarin Rat Snake, 129
Elaphe moellendorffi, Red-headed Rat Snake, 14, 127, 128
Elaphe obsoleta lindheimerii x quadrivittata, Texas Rat/Yellow Rat crossbreed, 118
Elaphe obsoleta lindheimerii, Texas Rat Snake, 118, 119, 127
Elaphe obsoleta obsoleta, Black Rat Snake, 5, 117, 128
Elaphe obsoleta quadrivittata, Yellow Rat Snake, 123, 126, 641
Elaphe obsoleta spiloides, Gray Rat Snake, 126
Elaphe obsoleta "williamsi", Gulf Hammock Rat Snake, 123

Elaphe radiata, Stripe-tailed Rat Snake, **122**
Elaphe rufodorsata, Rayed Rat Snake, **125**
Elaphe scalaris, Ladder Snake, **124**
Elaphe schrenckii, Manchurian Black Rat Snake, **124**
Elaphe [Senticolis] triaspis mutabilis, Mexican Green Snake, **121**
Elaphe [Senticolis] triaspis, Green Rat Snake, **116, 135**
Elaphe taeniura ridleyi, Ridley's Stripe-tailed Rat Snake, **121**
Elaphe taeniura yunnanensis, Yunnan Green Rat Snake, **135**
Elaphe vulpina vulpina, Western Fox Snake, **120**
Elaphe x Lampropeltis hybrid, **687**
Elapidae, 18
Elapomorphus bilineatus, Two-lined Burrowing Snake, **330**
Elapsoidea sundevalli boulengeri, Sundevall's Garter Snake, **414**
elegans eburnata, Arizona, Desert Glossy Snake, **89**
elegans philipi, Arizona, Painted Desert Glossy Snake, **90**
elegans, Arizona, Glossy Snake, **89**
Emerald Tree Boa, *Corallus caninus,* **9, 53, 54, 619**
Emydocephalus annulatus, Indonesian Turtle-headed Sea Snake, **457**
Endoparasites, 667
Enhydrina schistosa, Beaked Sea Snake, **458**
Enhydris chinensis, Chinese Water Snake, **309**
Enhydris jagerii, Jagor's Water Snake, **310**
Enhydris plumbea, Yellow-bellied Water Snake, **310**
Enulius flavitorques, Mexican Ground Snake, **331**
Environmental deficiencies, 664
enydris, Corallus, Garden Boa, **53, 54**
Epicrates angulifer, Cuban Boa, **670**
Epicrates cenchria, Rainbow Boa, **55**
Epicrates monensis granti, Virgin Islands Tree Boa, **56**
Epicrates monensis monensis, Mona Boa, **56**
Epicrates subflavus, Jamaican Boa, **55**
epinephalus, Liophis, Smooth Ground Snake, **82**
eques megalops, Thamnophis, Mexican Garter Snake, **300**
Eristicophis macmahonii, Leaf-nosed Viper, **490**
Erpeton tentaculatus, Tentacled Snake, **311**
Erycinae, 10
erythrogaster erythrogaster, Nerodia, Redbelly Water Snake, **273**
erythrogaster flavigaster, Nerodia, Yellowbelly Water Snake, **14, 269, 272**
Erythrolamprus aesculapii, Aesculapean False Coral Snake, **334–335, 336**
Erythrolamprus bizonus, Slender False Coral Snake, **332–333**

Erythrolamprus mimus, Honduran False Coral Snake, **331**
erytrogramma erytrogramma, Farancia, Rainbow Snake, **239, 240, 241**
Eryx colubrinus loveridgei, Sri Lankan Sand Boa, **72**
Eryx jaculus turcicus, Turkish Sand Boa, **72**
Eryx johni, Brown Sand Boa, **71**
Eryx tataricus, Giant Sand Boa, **71**
Eunectes murinus, Green Anaconda, **57, 58–59, 60**
Eunectes notaeus, Yellow Anaconda, **60**
European Asp, *Vipera aspis,* **492–493**
European Montpellier Snake, *Malpolon monspessulanus,* **351**
euryxanthus, Micruroides, Arizona Coral Snake, **417**
Evolution, 579
extenuatum, Stilosoma, Short-tailed Snake, **232–233**
Eyelash Viper, *Bothrops schlegeli,* **532, 533, 536, 537, 541, 543, 544, 545, 547, 645, 656**
Eyes, swollen, **659**
fallax, Telescopus, False Cat Snake, **386**
False Cat Snake, *Telescopus fallax,* **386**
False Coral Snake, *Anilius scytale,* **36–37**
False Fer-de-lance, *Xenodon rhabdocephalus,* **88**
False Smooth Snake, *Macroprotodon cucullatus,* **348**
Farancia abacura abacura, Eastern Mud Snake, **238, 239**
Farancia erytrogramma erytrogramma, Rainbow Snake, **239, 240, 241**
fasciata confluens, Nerodia, Broad-banded Water Snake, **269, 271**
fasciata fasciata, Nerodia, Banded Water Snake, **271, 275**
fasciata pictiventris, Nerodia, Florida Water Snake, **272, 274**
fasciatus, Bungarus, Banded Krait, **403, 406**
feae, Azemiops, Fea's Viper, **462**
Fea's Viper, *Azemiops feae,* **462**
Feeding, snakes's frequency of, 607
Fer-de-Lance, *Bothrops atrox,* **535**
Ficimia streckeri, Mexican Hooknose Snake, **337**
fieldi, Vipera [Pseudocerastes], Israeli Horned Viper, **491**
Fish, as a food item, 639
fiski, Lamprophis, Fisk's House Snake, **243**
Fisk's House Snake, *Lamprophis fiski,* **243**
flagellum flagellum, Masticophis, Eastern Coachwhip, **200**
flagellum piceus, Masticophis, Red Coachwhip, **201, 203**
flagellum testaceus, Masticophis, Western Coachwhip, **200, 202**
flagellum, Masticophis, Coachwhip, **199, 697**
Flat-nosed Pit Viper, *Trimeresurus puniceus,* **570**

719

flaviceps, Bungarus, Red-headed Krait, **407**
flavilata, Rhadinaea, Pine Woods Snake, **86**
flavitorques, Enulius, Mexican Ground Snake, **331**
Florida Brown Snake, *Storeria dekayi victa,* **282–283**
Florida Kingsnake, *Lampropeltis getula floridana,* **146, 171**
Florida Pine Snake, *Pituophis melanoleucus mugitus,* **212, 218, 617**
Florida Scarlet Snake, *Cemophora coccinea coccinea,* **90**
Florida Water Snake, *Nerodia fasciata pictiventris,* **272, 274**
Folklore, **690**
Food items, **635**
Force-feeding, **644**
Formosan Slug Snake, *Pareas formosensis,* **395**
formosensis, Pareas, Formosan Slug Snake, **395**
freminvillei, Stenorrhina, Rough-skinned Ground Snake, **382**
Fringed File Snake, *Mehelya crossi,* **248**
frontalis frontalis, Micrurus, Coral Cobra, **421**
frontalis, Prosymna, Northern Coppery Snake, **219**
fulgidus, Oxybelis, Green Vine Snake, **356**
fuliginosus, Boaedon, Common House Snake, **236, 237**
fulvius barbouri, Micrurus, Barbour's Coral Snake, **420**
fulvius tenere, Micrurus, Texas Coral Snake, **418, 424**
fulvius, Micrurus, Coral Snake, **418, 420, 422–423, 425**
Furina diadema, Red-naped Diadem Snake, **414**
Furnishings, for the terrarium, **629**
fusca, Boiga, Tan Tree Snake, **318**
fuscus, Aipysurus, Coastal Olive Sea Snake, **454**
gabonica rhinoceros, Bitis, Gaboon Viper, **476**
gabonica, Bitis, Rhinoceros Viper, **477, 479, 609, 657**
Gaboon Viper, *Bitis gabonica,* **477, 479, 609, 657**
Garden Boa, *Corallus enydris,* **53, 54**
Garden Tree Snake, *Boiga atriceps,* **317**
gerardi, Chilorhinophis, Tanganyikan Two-headed Snake, **397**
getula californiae, Lampropeltis, California Kingsnake, **142, 143, 146, 163, 170**
getula floridana, Lampropeltis, Florida Kingsnake, **146, 171**
getula getula, Lampropeltis, Eastern Chain Kingsnake, **172**
getula goini, Lampropeltis, Blotched Kingsnake, **142**
getula holbrooki, Lampropeltis, Speckled Kingsnake, **173**
getula niger, Lampropeltis, Black Kingsnake, **144–145, 174**
getula nigrita, Lampropeltis, Mexican Black Kingsnake, **166**
getula splendida, Lampropeltis, Desert Kingsnake, **169**
Giant Amazon Water Snake, *Hydrodynastes gigas,* **340, 341**
Giant Sand Boa, *Eryx tataricus,* **71**
gigas, Hydrodynastes, Giant Amazon Water Snake, **340, 341**
girondica, Coronella, Southern Smooth Snake, **106**
Glossy Crayfish Snake, *Regina rigida rigida,* **276**
Glossy Snake, *Arizona elegans,* **89**
Gloyd's Cantil, *Agkistrodon bilineatus howardgloydi,* **528**
Golden Flying Snake, *Chrysopelea ornata,* **319, 320, 321**
Golden-crowned Snake, *Cacophis squamulosus,* **408**
Gonyosoma oxycephalum, Red-tailed Rat Snake, **260**
gouldi, Unechis, Black-headed Snake, **448**
grahami, Regina, Graham's Crayfish Snake, **277**
grahamiae grahamiae, Salvadora, Mountain Patchnose Snake, **226, 227**
Graham's Crayfish Snake, *Regina grahami,* **277**
gramineus, Trimeresurus, Bamboo Pit Viper, **566**
Gravid, **676**
Gravidity, **676**
Gray Indigo Snake, *Drymarchon corais melanurus,* **109, 111**
Gray Rat Snake, *Elaphe obsoleta spiloides,* **126**
Gray-banded Kingsnake, *Lampropeltis alterna,* **138, 139, 140, 141, 168**
Gray-bellied Grass Snake, *Psammophylax rhombeatus,* **367, 368**
Grayia smithi, Smith's Water Snake, **260**
Great Lakes Bush Viper, *Atheris nitschei,* **468**
Great Plains Ground Snake, *Sonora semiannulata,* **375, 376–377, 379**
Great Plains Rat Snake, *Elaphe guttata emoryi,* **120**
Green Anaconda, *Eunectes murinus,* **57, 58–59, 60**
Green Bush Viper, *Atheris chloroechis,* **469**
Green Cat-eyed Snake, *Boiga cyanea,* **313**
Green Mamba, *Dendroaspis angusticeps,* **412**
Green Rat Snake, *Elaphe [Senticolis] triaspis,* **116, 135**
Green Tree Python, *Chondropython viridis,* **41, 42**
Green Vine Snake, *Ahaetulla prasina,* **15, 312**
Green Vine Snake, *Oxybelis fulgidus,* **356**
Ground Snake, *Sonora semiannulata,* **378, 379**
Guatemalan Milk Snake, *Lampropeltis*

720

triangulum abnorma, 177
Guatemalan Mussurana, *Clelia clelia,* 323
Guatemalan Palm Viper, *Bothrops [Bothriechis] aurifer,* 533
Guatemalan Racer, *Leptodrymus pulcherrimus,* 193
guerini, Phimophis, Panamanian Ground Racer, 361
Gulf Hammock Rat Snake, *Elaphe obsoleta "williamsi",* 123
guttata emoryi, Elaphe, Great Plains Rat Snake, 120
guttata guttata, Elaphe, Corn Snake, 117, 119, 129, 130, 131, 132–133, 581, 633
guttatus, Lamprophis, Spotted House Snake, 243
guttatus, Pseudechis, Spotted Black Snake, 441
Gyalopion canum, Western Hooknose Snake, 338
Gyalopion quadrangularis, Desert Hooknose Snake, 338
Habitat, of snakes, 648, 649
haemachatus, Hemachatus, Spitting Cobra, 415, 416
Haitian Longtail Snake, *Uromacer dorsalis,* 87
haje, Naja, Egyptian Cobra, 435
Half-Banded Hognose Snake, *Lystrophis semicinctus,* 83, 84
Half-banded Sea Krait, *Laticauda semifasciata,* 452
Half-ringed Cat Snake, *Telescopus semiannulatus,* 384, 385
hammondii, Thamnophis, Two-striped Garter Snake, 294
Handling, 652
hannah, Ophiophagus, King Cobra, 437, 438, 439, 440
Hapsidophrys lineata, Black-lined Green Snake, 138
Hearing, sense of, 598
Heat, snakes's balance of, 610
Heaters, aquarium, 624
Heaters, cable, 624
Heaters, ceramic, 623
Heaters, pad, 624
Heaters, tape, 624
Heaters, types of, 623
Heating, central, 624
Heating, equipment, 622, 625
Heating, products, 621
Heating, terraria, 620
helenae, Elaphe, Trinket Snake, 134
Helicops carinicauda, Common Keelback Snake, 261
Helicops leopardinus, Leopard Keelback Snake, 261
Helicops sp., Keelback Snake, 262–263
Hemachatus haemachatus, Spitting Cobra, 415, 416
Hemiaspis signata, Black-bellied Swamp Snake, 416

Hemirhagerrhis nototaenia, Bark Snake, 339
Heterodon nasicus kennerlyi, Mexican Hognose Snake, 77
Heterodon nasicus, Western Hognosed Snake, 76, 593, 607
Heterodon platirhinos, Eastern Hognose Snake, 12, 78–79, 80, 583, 634
Heterodon simus, Southern Hognose Snake, 77
Hibernation, 657
hippocrepis hippocrepis, Coluber, Horseshoe Racer, 98, 651
hispidus, Atheris, Laurent's Mountain Bush Viper, 468, 469
Hog-nosed Pit Viper, *Bothrops nasutus,* 542
Holding snake, proper method of, 651, 653
Homalopsinae, 13
Homo sapiens, 612, 621, 688
Homoroselaps lacteus, Southern Dwarf Garter Snake, 397
Honduran False Coral Snake, *Erythrolamprus mimus,* 331
Honduran Milk Snake, *Lampropeltis triangulum hondurensis,* 162, 183
Hook-nosed Beaked Snake, *Rhamphiophis oxyrhynchus,* 370, 371, 372
Hoplocephalus bitorquatus, Pale-headed Snake, 417
hoplogaster, Philothamnus, Common Green Bush Snake, 207
Horned Adder, *Bitis caudalis,* 471, 472–473, 474–475, 481, 482
Horned Sand Viper, *Cerastes cerastes,* 486
horridus atricaudatus, Crotalus, Canebrake Rattlesnake, 557
horridus horridus, Crotalus, Timber Rattlesnake, 5, 559, 699
Horseshoe Racer, *Coluber hippocrepis,* 98, 651
hotamboeia, Crotaphopeltis, Amboina Yellow-lipped Snake, 326
Huachuca Mountain Kingsnake, *Lampropeltis pyromelana woodini,* 176
Humidity, 626
humilis, Leptotyphlops, Western Blind Snake, 35
Hydrodynastes bicinctus, Blotched Amazon Water Snake, 339
Hydrodynastes gigas, Giant Amazon Water Snake, 340, 341
Hydrophiidae, 20
Hydrophiinae, 22
Hydrophis cyanocinctus, Blue-banded Sea Snake, 458, 459
Hydrophis inornatus, Soft-banded Sea Snake, 460
Hydrophis semperi, Lake Taal Banded Sea Snake, 459
Hydrophis sp., banded sea snake, 597
hypnale, Agkistrodon [Hypnale], Merrem's Hump-nosed Viper, 529, 530
Hypsiglena torquata, Night Snake, 341

Imantodes cenchoa, Blunt-headed Tree Snake, 342
Imantodes sp., Tree Snake, 343
Immobilizing snake, method of, 653
imperialis, Coniophanes, Royal Black-striped Snake, 324–325
Incubation, 679
Indian Krait, *Bungarus caeruleus*, 405, 406
Indo-Chinese Rat Snake, *Ptyas korros*, 222, 223
Indonesian Bronzeback, *Dendrelaphis pictus*, 107
Indonesian Cobra, *Naja naja sputatrix*, 430
Indonesian Turtle-headed Sea Snake, *Emydocephalus annulatus*, 457
Infections, bacterial, 668
Infections, protozoan, 668
Infectious stomatitis, 661, 668
Infra-red Lamps, 623
inornata, Bitis, Plain Mountain Adder, 480
inornata, Dasypeltis, South African Egg-eating Snake, 236, 603
inornatus, Hydrophis, Soft-banded Sea Snake, 460
intermedius caucasicus, Agkistrodon, Caspian Pit Viper, 528
Invertebrates, as a food item, 635
irregularis, Boiga, Brown Tree Snake, 314, 315
Israeli Horned Viper, *Vipera [Pseudocerastes] fieldi*, 491
jaculus turcicus, Eryx, Turkish Sand Boa, 72
jagerii, Enhydris, Jagor's Water Snake, 310
Jagor's Water Snake, *Enhydris jagerii*, 310
Jalisco Milk Snake, *Lampropeltis triangulum arcifera*, 163, 179
jallae, Psammophis, Rhodesian Sand Snake, 362
Jamaican Boa, *Epicrates subflavus*, 55
jamesoni, Dendroaspis, Jameson's Mamba, 21, 411
Jameson's Mamba, *Dendroaspis jamesoni*, 21, 411
Jan's Pine Snake, *Pituophis deppei jani*, 216
Jararaca, *Bothrops jararaca*, 539, 546
jararaca, Bothrops, Jararaca, 539, 546
johni, Eryx, Brown Sand Boa, 71
jugularis, Coluber, Large Whip Snake, 102
Jumping Viper, *Bothrops nummifer*, 538, 547
karelini mintonorum, Coluber, Minton's Spotted Mountain Racer, 102
kaznakovi, Vipera, Caucasus Viper, 505, 506
Keelback Snake, *Helicops* sp., 262–263
Keeled Slug Snake, *Pareas carinatus*, 394
Keeled Snake, *Pythonodipsas carinata*, 370
Keeled Water Snake, *Amphiesma* sp., 253
King Cobra, *Ophiophagus hannah*, 437, 438, 439, 440
kirtlandi capensis, Thelotornis, Cape Twig Snake, 387
korros, Ptyas, Indo-Chinese Rat Snake, 222, 223
Lachesis muta, Bushmaster, 560

lacteus, Homoroselaps, Southern Dwarf Garter Snake, 397
Ladder Snake, *Elaphe scalaris*, 124
Lake Taal Banded Sea Snake, *Hydrophis semperi*, 459
Lampropeltis alterna, Gray-banded Kingsnake, 138, 139, 140, 141, 168
Lampropeltis calligaster calligaster, Prairie Kingsnake, 166
Lampropeltis calligaster occipitolineata, South Florida Mole Kingsnake, 167
Lampropeltis calligaster rhombomaculata, Mole Snake, 167
Lampropeltis getula californiae, California Kingsnake, 142, 143, 146, 163, 170
Lampropeltis getula floridana, Florida Kingsnake, 146, 171
Lampropeltis getula getula, Eastern Chain Kingsnake, 172
Lampropeltis getula goini, Blotched Kingsnake, 142
Lampropeltis getula holbrooki, Speckled Kingsnake, 173
Lampropeltis getula niger, Black Kingsnake, 144–145, 174
Lampropeltis getula nigrita, Mexican Black Kingsnake, 166
Lampropeltis getula splendida, Desert Kingsnake, 169
Lampropeltis hybrid, 584
Lampropeltis mexicana "greeri", Durango Mountain Kingsnake, 147, 150, 151, 174
Lampropeltis mexicana "thayeri", Nuevo Leon Kingsnake, 148, 149
Lampropeltis mexicana, San Luis Potosi Kingsnake, 169
Lampropeltis pyromelana infralabialis, Utah Mountain Kingsnake, 175
Lampropeltis pyromelana knoblochi, Chihuahua Mountain Kingsnake, 162, 175
Lampropeltis pyromelana pyromelana, Arizona Mountain Kingsnake, 176
Lampropeltis pyromelana woodnini, Huachuca Mountain Kingsnake, 176
Lampropeltis ruthveni, Ruthven's Kingsnake, 177
Lampropeltis triangulum abnorma, Guatemalan Milk Snake, 177
Lampropeltis triangulum amaura, Louisiana Milk Snake, 178
Lampropeltis triangulum andesiana, Andean Milk Snake, 172
Lampropeltis triangulum annulata, Mexican Milk Snake, 153, 154–155, 178
Lampropeltis triangulum arcifera, Jalisco Milk Snake, 163, 179
Lampropeltis triangulum blanchardi, Blanchard's Milk Snake, 179
Lampropeltis triangulum campbelli, Pueblan Milk Snake, 153, 161, 180
Lampropeltis triangulum celaenops, New Mexico Milk Snake, 180

Lampropeltis triangulum conanti, Conant's Milk Snake, **161, 181**
Lampropeltis triangulum dixoni, Dixon's Milk Snake, **181**
Lampropeltis triangulum elapsoides, Scarlet Kingsnake, **157, 168, 669, 691**
Lampropeltis triangulum gaigeae, Black Milk Snake, **182**
Lampropeltis triangulum gentilis, Central Plains Milk Snake, **182**
Lampropeltis triangulum hondurensis, Honduran Milk Snake, **162, 183**
Lampropeltis triangulum micropholis, Ecuadorian Milk Snake, **183**
Lampropeltis triangulum multistrata, Pale Milk Snake, **184**
Lampropeltis triangulum nelsoni, Nelson's Milk Snake, **173**
Lampropeltis triangulum oligozona, Pacific Central American Milk Snake, **184**
Lampropeltis triangulum polyzona, Atlantic Central American Milk Snake, **185**
Lampropeltis triangulum sinaloae, Sinaloan Milk Snake, **158, 159, 185, 615**
Lampropeltis triangulum smithi, Smith's Milk Snake, **186**
Lampropeltis triangulum stuarti, Stuart's Milk Snake, **186, 577**
Lampropeltis triangulum syspila, Red Milk Snake, **160, 187**
Lampropeltis triangulum taylori, Utah Milk Snake, **187**
Lampropeltis triangulum "temporalis", Coastal Plains Milk Snake, **156**
Lampropeltis triangulum triangulum, Eastern Milk Snake, **152, 188, 658**
Lampropeltis x Elaphe hybrid, **687**
Lampropeltis zonata agalmae, San Pedro Mountain Kingsnake, **188**
Lampropeltis zonata herrerae, Todos Santos Island Kingsnake, **189**
Lampropeltis zonata multicincta, Sierra Mountain Kingsnake, **189**
Lampropeltis zonata multifasciata, Coastal Mountain Kingsnake, **190, 192**
Lampropeltis zonata parvirubra, San Bernardino Mountain Kingsnake, **190**
Lampropeltis zonata pulchra, San Diego Mountain Kingsnake, **191**
Lampropeltis zonata zonata, St. Helena Mountain Kingsnake, **191**
Lampropeltis zonata, Mountain Kingsnake, **164–165, 192**
Lamprophis aurora, Aurora House Snake, **242**
Lamprophis fiski, Fisk's House Snake, **243**
Lamprophis guttatus, Spotted House Snake, **243**
Large Coral Snake, *Micrurus lemniscatus,* **424**
Large Whip Snake, *Coluber jugularis,* **102**
Large-eyed Snake, *Telescopus dhara,* **384**
Latefi's Viper, *Vipera latifi,* **507**
lateralis euryxanthus, Masticophis, Alameda Striped Racer, **201**
lateralis lateralis, Masticophis, California Striped Racer, **203**
lateralis, Bothrops, Yellow-lipped Palm Viper, **543, 545**
Laticauda colubrina, Yellow-lipped Sea Krait, **451, 453**
Laticauda semifasciata, Half-banded Sea Krait, **452**
Laticauda sp., Sea Krait, **451, 452**
Laticaudinae, 21
latifi, Vipera, Latefi's Viper, **507**
Laurent's Mountain Bush Viper, *Atheris hispidus,* **468, 469**
Leaf-nosed Viper, *Eristicophis macmahonii,* **490**
lebetina, Vipera, Levantine Viper, **495, 507**
lecontei antonii, Rhinocheilus, Mexican Longnose Snake, **225**
lecontei tessellatus, Rhinocheilus, Texas Longnose Snake, **224, 225**
Legless lizard, **590**
leightoni namibensis, Psammophis, Namib Desert Sand Snake, **363**
leightoni trinasalis, Psammophis, Leighton's Sand Snake, **366**
Leighton's Sand Snake, *Psammophis leightoni trinasalis,* **366**
leithi, Psammophis, Pakistan Ribbon Snake, **364**
lemniscatus, Micrurus, Large Coral Snake, **424**
Leopard Keelback Snake, *Helicops leopardinus,* **261**
leopardinus, Helicops, Leopard Keelback Snake, **261**
lepidus klauberi, Crotalus, Banded Rock Rattlesnake, **556**
Leptodeira annulata, Ringed Cat-eyed Snake, **346**
Leptodeira nigrofasciata, Black-banded Cat-eyed Snake, **346**
Leptodeira septentrionalis septentrionalis, Northern Cat-eyed Snake, **344–345, 348**
Leptodeira septentrionalis, Cat-eyed Snake, **347**
Leptodrymus pulcherrimus, Guatemalan Racer, **193**
Leptophis ahaetulla, Parrot Snake, **195, 196**
Leptophis diplotrophis, Sunrise Parrot Snake, **194**
Leptophis mexicanus, Mexican Parrot Snake, **193, 194**
Leptotyphlopidae, 6
Leptotyphlops dulcis dissectus, New Mexico Blind Snake, **35**
Leptotyphlops dulcis, Texas Blind Snake, **7**
Leptotyphlops humilis, Western Blind Snake, **35**
Levantine Viper, *Vipera lebetina,* **495, 507**
Liasis albertisii, D'Albert's Python, **43**
Liasis boa, Ringed Python, **44**
Liasis papuanus, Papuan Python, **44**

723

Lichanura trivirgata gracia, Desert Rosy Boa, 61
Lichanura trivirgata "myriolepis", Central Baja Rosy Boa, 62
Lichanura trivirgata trivirgata, Mexican Rosy Boa, 61, 62
Lighting, 624
Lighting, broad-spectrum, 625
Lighting, halogen and mercury vapor, 626
lineata, Hapsidophrys, Black-lined Green Snake, 138
lineatum lineatum, Tropiclonion, Common Lined Snake, 304–305
lineatum texanum, Tropiclonion, Texas Lined Snake, 302–303
lineatus, Conopsis, Queretaro Ground Snake, 75
lineolatus, Psammophis, Steppe Ribbon Snake, 365
Lining, cage, 632
Liophis anomala, Broken Ground Snake, 82
Liophis epinephalus, Smooth Ground Snake, 82
Liophis reginae, Royal Ground Snake, 81
Liophis sp., South American Ground Snake, 81
Liopholidophis sexlineatus, Six-lined Snake, 244
Liopholidophis stumpffi, Stumpff's Grass Snake, 244
Liotyphlops albirostris, Central American Blind Snake, 34
Little Spotted Snake, *Denisonia punctata,* 413
Locomotion, 594
Locusts, as a food item, 635
Long-nosed Vine Snake, *Ahaetulla nasuta,* 312
longissima, Elaphe, Aesculapian Snake, 701–702
Louisiana Milk Snake, *Lampropeltis triangulum amaura,* 178
Louisiana Pine Snake, *Pituophis ruthveni,* 213
Lowland Viper, *Atheris superciliaris,* 470
Loxoceminae, 8
Loxocemus bicolor, Mexican Burrowing Python, 38
lubricus, Aspidelaps, Cape Coral Cobra, 400, 401
lutrix, Duberria, Abyssinia Slug-eater, 114
Lycodon aulicus, Common Wolf Snake, 246
Lycodon striatus bicolor, Shaw's Two-colored Wolf Snake, 245
Lycodon subcinctus, Banded Wolf Snake, 246
Lycodonomorphus whytii, White-lipped Water Snake, 247
Lycodontinae, 12
Lycophidion capense, Cape Wolf Snake, 247
Lyre Snake, *Trimorphodon biscutatus,* 627
Lystrophis dorbignyi, Dorbigny's Hognose Snake, 83, 84
Lystrophis semicinctus, Half-Banded Hognose Snake, 83, 84

Lytorhynchus maynardi, Maynard's Awl-headed Snake, 197
Lytorhynchus ridgewayi, Afghan Awl-headed Snake, 197
macclellandi, Calliophis, MacClelland's Oriental Coral Snake, 409
MacClelland's Oriental Coral Snake, *Calliophis macclellandi,* 409
macgregori, Trimeresurus, MacGregor's Pit Viper, 566
MacGregor's Pit Viper, *Trimeresurus macgregori,* 566
macmahonii, Eristicophis, Leaf-nosed Viper, 490
Macroprotodon cucullatus, False Smooth Snake, 348
maculatus, Causus, Spotted Adder, 511
Madagascan Blunt-nosed Snake, *Madagascarophis colubrinus,* 349, 350
Madagascan False Racer, *Mimophis madagascariensis,* 352
Madagascar Ground Boa, *Acrantophis madagascariensis,* 50
Madagascar Tree Boa, *Sanzania madagascariensis,* 63
madagascariensis, Acrantophis, Madagascar Ground Boa, 50
madagascariensis, Mimophis, Madagascan False Racer, 352
madagascariensis, Sanzania, Madagascar Tree Boa, 63
Madagascarophis colubrinus, Madagascan Blunt-nosed Snake, 349, 350
Mahafalan False Racer, *Mimophis mahafalensis,* 352
mahafalensis, Mimophis, Mahafalan False Racer, 352
Malayan Blood Python, *Python curtus brongersmai,* 48
Malayan Bush Snake, *Rhabdophis subminiatus,* 278
Malayan Krait, *Bungarus candidus,* 407, 408
Malaysian Moccasin, *Agkistrodon rhodostoma,* 529
Malpolon moilensis, Atlas Mountain Montpellier Snake, 18, 350, 351
Malpolon monspessulanus, European Montpellier Snake, 351
Mammals, as a food item, 643
Manchurian Black Rat Snake, *Elaphe schrenckii,* 124
Mandarin Rat Snake, *Elaphe mandarina,* 129
mandarina, Elaphe, Mandarin Rat Snake, 129
Mangrove Pit Viper, *Trimeresurus purpureomaculatus,* 564, 565, 568
Mangrove Snake, *Boiga dendrophila,* 314, 316, 653
Many Spotted Tree Snake, *Boiga multomaculata,* 318
Many-banded Krait, *Bungarus multicinctus,* 404, 405
Many-horned Adder, *Bitis cornuta,* 484

marcianus, Thamnophis, Checkered Garter Snake, **289, 295**
margaritiferus, Drymobius, Mexican Speckled Snake, **112–113, 114**
margaritophorus, Pareas, White-spotted Slug Snake, **394, 395**
Maricopa Leafnose Snake, *Phyllorhynchus browni lucidus,* **209**
Massasauga, *Sistrurus catenatus,* **561**
Masticophis bilineatus, Sonoran Whipsnake, **198, 204**
Masticophis flagellum flagellum, Eastern Coachwhip, **200**
Masticophis flagellum piceus, Red Coachwhip, **201, 203**
Masticophis flagellum testaceus, Western Coachwhip, **200, 202**
Masticophis flagellum, Coachwhip, **199, 697**
Masticophis lateralis euryxanthus, Alameda Striped Racer, **201**
Masticophis lateralis lateralis, California Striped Racer, **203**
Masticophis taeniatus schotti, Schott's Whipsnake, **202**
Mastigodryas bifossatus, Venezuelan Tropical Racer, **204**
maura, Natrix, Viperine Water Snake, **265, 266–267**
maynardi, Lytorhynchus, Maynard's Awl-headed Snake, **197**
Maynard's Awl-headed Snake, *Lytorhynchus maynardi,* **197**
Mealworms, as a food item, **636**
mechovi inornatus, Xenocalamus, Ornate Quill-nosed Snake, **391**
Mehelya capensis, Cape File Snake, **248**
Mehelya crossi, Fringed File Snake, **248**
meixcana "thayeri", Lampropeltis, Nuevo Leon Kingsnake, **148**
melanocephalus, Aspidites, Black-headed Python, **39**
melanocephalus, Rhynchocalamus, Asian Black-headed Snake, **252**
melanoleuca, Naja, Black-lipped Cobra, **432, 434**
melanoleucus lodingi, Pituophis, Black Pine Snake, **217**
melanoleucus melanoleucus, Pituophis, Northern Pine Snake, **210, 212, 213, 214**
melanoleucus mugitus, Pituophis, Florida Pine Snake, **212, 218, 617**
melanurus melanurus, Tropidophis, Cuban Black-tailed Dwarf Boa, **64**
merremi, Xenodon, Sapera, **88**
Merrem's Hump-nosed Viper, *Agkistrodon [Hypnale] hypnale,* **529, 530**
Mexican Black Kingsnake, *Lampropeltis getula nigrita,* **166**
Mexican Burrowing Python, *Loxocemus bicolor,* **38**
Mexican Bush Racer, *Oxyrhopus petolarius,* **357, 358–359**

Mexican Garter Snake, *Thamnophis eques megalops,* **300**
Mexican Green Snake, *Elaphe [Senticolis] triaspis mutabilis,* **121**
Mexican Ground Snake, *Enulius flavitorques,* **331**
Mexican Hognose Snake, *Heterodon nasicus kennerlyi,* **77**
Mexican Hooknose Snake, *Ficimia streckeri,* **337**
Mexican Longnose Snake, *Rhinocheilus lecontei antonii,* **225**
Mexican Milk Snake, *Lampropeltis triangulum annulata,* **153, 154–155, 178**
Mexican Parrot Snake, *Leptophis mexicanus,* **193, 194**
Mexican Patchnose Snake, *Salvadora mexicanum,* **226**
Mexican Pine Snake, *Pituophis deppei deppei,* **216**
Mexican Rosy Boa, *Lichanura trivirgata trivirgata,* **61, 62**
Mexican Speckled Snake, *Drymobius margaritiferus,* **112–113, 114**
mexicana "greeri", Lampropeltis, Durango Mountain Kingsnake, **147, 150, 151, 174**
mexicana "thayeri", Lampropeltis, Nuevo Leon Kingsnake, **149**
mexicana, Lampropeltis, San Luis Potosi Kingsnake, **169**
mexicanum, Salvadora, Mexican Patchnose Snake, **226**
mexicanus, Leptophis, Mexican Parrot Snake, **193, 194**
Michoacan Ground Snake, *Sonora michoacanensis,* **380–381**
michoacanensis, Sonora, Michoacan Ground Snake, **380–381**
Micruroides euryxanthus, Arizona Coral Snake, **417**
Micrurus corallinus corallinus, Common Coral Snake, **421**
Micrurus dumerili, Dumeril's Coral Snake, **419**
Micrurus frontalis frontalis, Coral Cobra, **421**
Micrurus fulvius barbouri, Barbour's Coral Snake, **420**
Micrurus fulvius tenere, Texas Coral Snake, **418, 424**
Micrurus fulvius, Coral Snake, **418, 420, 422–423, 425**
Micrurus lemniscatus, Large Coral Snake, **424**
Micrurus nigrocinctus nigrocinctus, Common Black-banded Coral Snake, **425**
Micrurus nigrocinctus, Black-banded Coral Snake, **419**
miliarius miliarius, Sistrurus, Carolina Pygmy Rattlesnake, **562**
miliarius, Sistrurus, Pygmy Rattlesnake, **563, 600**
Milking, of a rattlesnake, **699**
Mimophis madagascariensis, Madagascan False Racer, **352**

725

Mimophis mahafalensis, Mahafalan False Racer, **352**
mimus, Erythrolamprus, Honduran False Coral Snake, **331**
Minton's Spotted Mountain Racer, *Coluber karelini mintonorum,* **102**
Mississippi Green Water Snake, *Nerodia cyclopion,* **270**
Misting, captive boas, **619**
mitchelli pyrrhus, Crotalus, Southwestern Speckled Rattlesnake, **550**
mitchelli, Crotalus, Speckled Rattlesnake, **673**
Mites, **666**
Mock Viper, *Psammodynastes pulverulentus,* **362**
moellendorffi, Elaphe, Red-headed Rat Snake, **14, 127, 128**
moilensis, Malpolon, Atlas Mountain Montpellier Snake, **18, 350, 351**
Mole Snake, *Lampropeltis calligaster rhombomaculata,* **167**
molossus, Crotalus, Northern Black-tailed Rattlesnake, **550**
Molslang, *Pseudaspis cana,* **220**
Mona Boa, *Epicrates monensis monensis,* **56**
monensis granti, Epicrates, Virgin Islands Tree Boa, **56**
monensis monensis, Epicrates, Mona Boa, **56**
Monocled Cobra, *Naja naja kaouthia,* **428**
monspessulanus, Malpolon, European Montpellier Snake, **351**
moojeni, Bothrops, Caissaca, **546**
Morelia spilotes variegata, Carpet Python, **45**
mossambica pallida, Naja, Pallid Spitting Cobra, **431**
mossambica, Naja, Spitting Cobra, **433**
Mountain Adder, *Bitis atropos,* **479**
Mountain Kingsnake, *Lampropeltis zonata,* **164–165, 192**
Mountain Patchnose Snake, *Salvadora grahamiae grahamiae,* **226, 227**
Mountain Racer, *Coluber ravergieri,* **98, 100, 101**
Mouth rot, **661, 668**
mucosus, Ptyas, Oriental Rat Snake, **222**
Mulga Snake, *Pseudechis australis,* **442–443**
multicinctus, Bungarus, Many-banded Krait, **404, 405**
multimaculata, Amplorhinus, Cape Many-spotted Snake, **313**
multomaculata, Boiga, Many Spotted Tree Snake, **318**
murinus, Eunectes, Green Anaconda, **57, 58–59, 60**
muta, Lachesis, Bushmaster, **560**
Naja haje, Egyptian Cobra, **435**
naja kaouthia, Naja, Monocled Cobra, **428**
Naja melanoleuca, Black-lipped Cobra, **432, 434**
Naja mossambica pallida, Pallid Spitting Cobra, **431**
Naja mossambica, Spitting Cobra, **433**

Naja naja kaouthia, Monocled Cobra, **428**
Naja naja philippinenesis, Philippine Cobra, **430**
Naja naja sputatrix, Indonesian Cobra, **430**
Naja naja sumatrana, Sumatran Cobra, **431**
Naja naja, Common Cobra, **426, 432**
Naja nigricollis, Black-necked Cobra, **427, 429**
Naja nivea, Cape Cobra, **433**
Naja oxiana, Central Asian Cobra, **429**
naja philippinenesis, Naja, Philippine Cobra, **430**
naja sputatrix, Naja, Indonesian Cobra, **430**
naja sumatrana, Naja, Sumatran Cobra, **431**
naja, Naja, Common Cobra, **426, 432**
najadum dahli, Coluber, Dahl's Whipsnake, **101**
Namib Desert Sand Snake, *Psammophis leightoni namibensis,* **363**
Namib Tiger Snake, *Telescopus beetzii,* **386**
nasicornis, Bitis, Rhinoceros Viper, **476, 478**
nasicus kennerlyi, Heterodon, Mexican Hognose Snake, **77**
nasicus, Heterodon, Western Hognosed Snake, **76, 593, 607**
nasuta, Ahaetulla, Long-nosed Vine Snake, **312**
nasutus, Bothrops, Hog-nosed Pit Viper, **542**
Natricinae, **13**
Natriciteres olivacea, Banded Olive Snake, **264**
natrix helvetica, Natrix, Yellow Grass Snake, **268**
Natrix maura, Viperine Water Snake, **265, 266–267**
Natrix natrix helvetica, Yellow Grass Snake, **268**
Natrix tessellata, Tesselated Water Snake, **264, 268**
nebulata, Sibon, Central American Snail-eater, **393**
Nelson's Milk Snake, *Lampropeltis triangulum nelsoni,* **173**
Neotropical Rattlesnake, *Crotalus durissus,* **692**
Nerodia clarkii, Salt Marsh Snake, **270, 273**
Nerodia cyclopion, Mississippi Green Water Snake, **270**
Nerodia erythrogaster erythrogaster, Redbelly Water Snake, **273**
Nerodia erythrogaster flavigaster, Yellowbelly Water Snake, **14, 269, 272**
Nerodia fasciata confluens, Broad-banded Water Snake, **269, 271**
Nerodia fasciata fasciata, Banded Water Snake, **271, 275**
Nerodia fasciata pictiventris, Florida Water Snake, **272, 274**
Nerodia rhombifer rhombifer, Diamondback Water Snake, **274**
Nerodia sipedon, Northern Water Snake, **628, 637**
Neuweid's False Boa, *Pseudoboa neuwiedi,*

369
neuwiedi, Bothrops, Neuwied's Pit Viper, **544**
neuwiedi, Pseudoboa, Neuwied's False Boa, 369
Neuwied's Pit Viper, *Bothrops neuwiedi,* **544**
New Caledonian Olive Sea Snake, *Aipysurus apraefrontalis,* **454**
New Guinea Viper Boa, *Candoia aspera,* **66, 67, 622**
New Guinean Keeled Water Snake, *Amphiesma stolata,* **253, 586**
New Mexico Blind Snake, *Leptotyphlops dulcis dissectus,* **35**
New Mexico Milk Snake, *Lampropeltis triangulum celaenops,* **180**
Night Snake, *Hypsiglena torquata,* **341**
nigrescens, Cryptophis, Eastern Small-eyed Snake, **410**
nigrescens, Ramphotyphlops, Australian Blind Worm Snake, **33**
nigriceps, Tantilla, Plains Blackheaded Snake, **382, 383**
nigricollis, Naja, Black-necked Cobra, **427, 429**
nigrocinctus nigrocinctus, Micrurus, Common Black-banded Coral Snake, **425**
nigrocinctus, Micrurus, Black-banded Coral Snake, **419**
nigrofasciata, Leptodeira, Black-banded Cat-eyed Snake, **346**
nigrostriatus, Unechis, Black-striped Snake, **447**
Ninia sebae, Ring-necked Coffee Snake, **249**
nitschei, Atheris, Great Lakes Bush Viper, **468**
nivea, Naja, Cape Cobra, **433**
Northern Black Racer, *Coluber constrictor constrictor,* **103, 104**
Northern Black-tailed Rattlesnake, *Crotalus molossus,* **550**
Northern Cat-eyed Snake, *Leptodeira septentrionalis septentrionalis,* **344–345, 348**
Northern Copperhead, *Agkistrodon contortrix mokasen,* **522, 523, 524**
Northern Coppery Snake, *Prosymna frontalis,* **219**
Northern Pacific Rattlesnake, *Crotalus viridis oreganus,* **549, 557**
Northern Pine Snake, *Pituophis melanoleucus melanoleucus,* **210, 212, 213, 214**
Northern Scarlet Snake, *Cemophora coccinea copei,* **91**
Northern Tree Snake, *Dendrelaphis calligaster,* **108**
Northern Viper, *Vipera berus,* **23, 498–499, 504, 505, 508**
Northern Water Snake, *Nerodia sipedon,* **628, 637**
Northwestern Garter Snake, *Thamnophis ordinoides,* **300**
notaeus, Eunectes, Yellow Anaconda, **60**
Notechis scutatus, Tiger Snake, **435, 436**
nototaenia, Hemirhagerrhis, Bark Snake, **339**

Nuevo Leon Kingsnake, *Lampropeltis mexicana "thayeri",* **148, 149**
nummifer, Bothrops, Jumping Viper, **538, 547**
Nutritional problems, **664**
obsoleta lindheimerii x quadrivittata, Elaphe, Texas Rat/Yellow Rat crossbreed, **118**
obsoleta lindheimerii, Elaphe, Texas Rat Snake, **118, 119, 127**
obsoleta obsoleta, Elaphe, Black Rat Snake, **5, 117, 128**
obsoleta quadrivittata, Elaphe, Yellow Rat Snake, **123, 126, 641**
obsoleta spiloides, Elaphe, Gray Rat Snake, **126**
obsoleta "williamsi", Elaphe, Gulf Hammock Rat Snake, **123**
occipitalis klauberi, Chionactis, Tucson Shovelnose Snake, **97**
occipitalis, Chionactis, Western Shovelnose Snake, **96**
occipitolutea, Clelia, Yellow-headed Mussurana, **322**
occipitomaculata, Storeria, Redbelly Snake, **280, 281**
ocellata, Boiga, Ocellated Tree Snake, **17, 315**
Ocellated Night Viper, *Tomodon ocellatus,* **388**
Ocellated Tree Snake, *Boiga ocellata,* **17, 315**
ocellatus, Tomodon, Ocellated Night Viper, **388**
Oligodon taeniatus, Variegated Kukri Snake, **249**
Oligodon taeniolatus, Streaked Kukri Snake, **250–251**
olivacea, Natriciteres, Banded Olive Snake, **264**
Olive Sea Snake, *Aipysurus* sp., **455**
Opheodrys aestivus, Rough Green Snake, **206**
Opheodrys vernalis, Smooth Green Snake, **205, 207**
Ophiophagus hannah, King Cobra, **437, 438, 439, 440**
ophryomegas, Bothrops, Bocourt's Lance-headed Snake, **540**
Opisthoglyphic, **604**
ordinoides, Thamnophis, Northwestern Garter Snake, **300**
Oriental Rat Snake, *Ptyas mucosus,* **222**
ornata, Chrysopelea, Golden Flying Snake, **319, 320, 321**
Ornate Quill-nosed Snake, *Xenocalamus mechovi inornatus,* **391**
Ottoman Viper, *Vipera xanthina,* **494, 496–497**
Oviparous, **676**
Ovoviviparous, **676**
oxiana, Naja, Central Asian Cobra, **429**
Oxybelis spp., Green Vine Snake, **356**
Oxybelis aeneus, American Vine Snake, **353**
Oxybelis argenteus, Silver Vine Snake, **354–355**
Oxybelis brevirostris, Short-nosed Vine Snake, **353**
Oxybelis fulgidus, Green Vine Snake, **356**

Oxybelis, Green Vine Snake, 356
oxycephalum, *Gonyosoma*, Red-tailed Rat Snake, 260
Oxyrhopus petolarius, Mexican Bush Racer, 357, 358–359
Oxyrhopus rhombifer, Argentinian Bush Racer, 357, 360
Oxyrhopus trigeminus, Brazilian Bush Racer, 360
oxyrhynchus, *Rhamphiophis*, Hook-nosed Beaked Snake, 370, 371, 372
Oxyuranus scutellatus, Taipan, 440, 441
Pacific Central American Milk Snake, *Lampropeltis triangulum oligozona*, 184
Pacific Gopher Snake, *Pituophis catenifer catenifer*, 211
Pacific Ringneck Snake, *Diadophis punctatus amabilis*, 259
Painted Desert Glossy Snake, *Arizona elegans philipi*, 90
Pakistan Ribbon Snake, *Psammophis leithi*, 364
palaestinae, *Vipera*, Palestine Viper, 502–503
palarostris, *Chionactis*, Sonoran Shovelnose Snake, 96, 97
Pale Milk Snake, *Lampropeltis triangulum multistrata*, 184
Pale-headed Snake, *Hoplocephalus bitorquatus*, 417
Palestine Viper, *Vipera palaestinae*, 502–503
Pallid Spitting Cobra, *Naja mossambica pallida*, 431
Palm Viper, *Bothrops [Bothriechis] bilineatus*, 531
Panamanian Ground Racer, *Phimophis guerini*, 361
Papuan Python, *Liasis papuanus*, 44
papuanus, *Liasis*, Papuan Python, 44
Pareas carinatus, Keeled Slug Snake, 394
Pareas formosensis, Formosan Slug Snake, 395
Pareas margaritophorus, White-spotted Slug Snake, 394, 395
Pareinae, 16
Parrot Snake, *Leptophis ahaetulla*, 195, 196
Pastimes, peripheral, 706
Patagonian Pit Viper, *Bothrops ammodytoides*, 534
Pelamis platurus, Yellow-bellied Sea Snake, 460, 461
Peninsula Ribbon Snake, *Thamnophis sauritus sackeni*, 296–297
peringueyi, *Bitis*, Dwarf Puff Adder, 481, 483
peroni, *Acalyptophis*, Australian Coral Reef Snake, 453
Pet trade, 705
petolarius, *Oxyrhopus*, Mexican Bush Racer, 357, 358–359
Philippine Cobra, *Naja naja philippinenesis*, 430
phillipsii, *Psammophis*, Phillips's Sand Snake, 366

Phillips's Sand Snake, *Psammophis phillipsii*, 366
Philodryas baroni, Argentinian Green Snake, 85
Philodryas psammophideus, Bolivian Green Snake, 85
Philothamnus hoplogaster, Common Green Bush Snake, 207
Philothamnus semivariegatus, Spotted Bush Snake, 208
Phimophis guerini, Panamanian Ground Racer, 361
Phyllorhynchus browni lucidus, Maricopa Leafnose Snake, 209
Phyllorhynchus browni, Saddled Leafnose Snake, 209
Phyllorhynchus decurtatus perkinsi, Western Leafnose Snake, 208
Phyllorhynchus decurtatus, Spotted Leafnose Snake, 210
Physiology, of snakes, 591
piceivittis, *Coniophanes*, Double Black-striped Snake, 326
pictus, *Dendrelaphis*, Indonesian Bronzeback, 107
Pine Woods Snake, *Rhadinaea flavilata*, 86
piscator, *Xenochrophis*, Checkered Keelback, 308
piscivorus leucostoma, *Agkistrodon*, Western Cottonmouth, 516–517
piscivorus, *Agkistrodon*, Cottonmouth, 515, 520, 523
Pituophis catenifer affinis, Sonoran Gopher Snake, 215, 631
Pituophis catenifer catenifer, Pacific Gopher Snake, 211
Pituophis deppei deppei, Mexican Pine Snake, 216
Pituophis deppei jani, Jan's Pine Snake, 216
Pituophis melanoleucus lodingi, Black Pine Snake, 217
Pituophis melanoleucus melanoleucus, Northern Pine Snake, 210, 212, 213, 214
Pituophis melanoleucus mugitus, Florida Pine Snake, 212, 218, 617
Pituophis ruthveni, Louisiana Pine Snake, 213
Pituophis sayi, Bullsnake, 215, 218, 680–681
Plain Mountain Adder, *Bitis inornata*, 480
Plains Blackheaded Snake, *Tantilla nigriceps*, 382, 383
Plains Garter Snake, *Thamnophis radix*, 288, 293, 614
Plants, 632
platirhinos, *Heterodon*, Eastern Hognose Snake, 12, 78–79, 80, 583, 634
platurus, *Pelamis*, Yellow-bellied Sea Snake, 460, 461
Pliocercus sp., Big-scaled False Coral Snake, 275
plumbea, *Enhydris*, Yellow-bellied Water Snake, 310
poecilonotus, *Pseustes*, Amazonian Liar, 221

polylepis, Dendroaspis, Black Mamba, **412**
popeorum, Trimeresurus, Pope's Pit Viper, **567, 568**
Pope's Pit Viper, *Trimeresurus popeorum,* **567, 568**
porphyriacus, Pseudechis, Red-bellied Black Snake, **444–445**
Prairie Kingsnake, *Lampropeltis calligaster calligaster,* **166**
Prairie Ringneck Snake, *Diadophis punctatus arnyi,* **258**
prasina, Ahaetulla, Green Vine Snake, **15, 312**
Prey, snakes's method of capturing and killing, 605
Prey, snakes's swallowing of, 606
Proctodeum, 609
Prosymna bivittata, Two-striped Shovel-nosed Snake, **219**
Prosymna frontalis, Northern Coppery Snake, **219**
Proteroglyphic, 604
proximus proximus, Thamnophis, Western Ribbon Snake, **292**
proximus rubrilineatus, Thamnophis, Redstripe Ribbon Snake, **292**
Psammodynastes pulverulentus, Mock Viper, **362**
psammophideus, Philodryas, Bolivian Green Snake, **85**
Psammophis jallae, Rhodesian Sand Snake, **362**
Psammophis leightoni namibensis, Namib Desert Sand Snake, **363**
Psammophis leightoni trinasalis, Leighton's Sand Snake, **366**
Psammophis leithi, Pakistan Ribbon Snake, **364**
Psammophis lineolatus, Steppe Ribbon Snake, **365**
Psammophis phillipsii, Phillips's Sand Snake, **366**
Psammophis sp., Sand Snake, **363**
psammophis, Demansia, Yellow-faced Whip Snake, **411**
Psammophylax rhombeatus, Gray-bellied Grass Snake, **367, 368**
Psammophylax tritaeniatus, White-bellied Grass Snake, **17, 368, 369**
Pseudaspis cana, Molslang, **220**
Pseudechis australis, Mulga Snake, **442–443**
Pseudechis guttatus, Spotted Black Snake, **441**
Pseudechis porphyriacus, Red-bellied Black Snake, **444–445**
Pseudoboa neuwiedi, Neuweid's False Boa, **369**
Pseudonaja textilis, Eastern Brown Snake, **446**
Pseustes poecilonotus, Amazonian Liar, **221**
Pseustes sulphureus, Brazilian Liar, **221**
Ptyas korros, Indo-Chinese Rat Snake, **222, 223**
Ptyas mucosus, Oriental Rat Snake, **222**

Pueblan Milk Snake, *Lampropeltis triangulum campbelli,* **153, 161, 180**
pulcherrimus, Leptodrymus, Guatemalan Racer, **193**
pullatus, Spilotes, Black and Yellow Rat Snake, **231**
pulverulentus, Dipsadoboa, Dusky Tropical Cat-eyed Snake, **327**
pulverulentus, Psammodynastes, Mock Viper, **362**
punctata, Denisonia, Little Spotted Snake, **413**
punctatus amabilis, Diadophis, Pacific Ringneck Snake, **259**
punctatus arnyi, Diadophis, Prairie Ringneck Snake, **258**
punctatus pulchellus, Diadophis, Coralbelly Ringneck Snake, **257**
punctatus punctatus, Diadophis, Southern Ringneck Snake, **257**
punctatus regalis, Diadophis, Regal Ringneck Snake, **256**
punctatus, Diadophis, Ringneck Snake, **255, 256**
punctulatus, Dendrelaphis, Common Tree Snake, **107**
puniceus, Trimeresurus, Flat-nosed Pit Viper, **570**
purpureomaculatus, Trimeresurus, Mangrove Pit Viper, **564, 565, 568**
pygaea cyclas, Seminatrix, South Florida Swamp Snake, **279**
pygaea, Seminatrix, Black Swamp Snake, **279**
Pygmy Rattlesnake, *Sistrurus miliarius,* **563, 600**
pyromelana infralabialis, Lampropeltis, Utah Mountain Kingsnake, **175**
pyromelana knoblochi, Lampropeltis, Chihuahua Mountain Kingsnake, **162, 175**
pyromelana pyromelana, Lampropeltis, Arizona Mountain Kingsnake, **176**
pyromelana woodnini, Lampropeltis, Huachuca Mountain Kingsnake, **176**
pyrrhus, Acanthophis, Australian Death Adder, **399**
Python anchietae, Angola Python, **49**
Python boeleni, Boelen's Python, **9, 46**
Python curtus brongersmai, Malayan Blood Python, **48**
Python curtus, Blood Python, **47, 659**
Python regius, Ball Python, **48, 612, 661, 669**
Python reticulatus, Reticulated Python, **46, 662**
Python timorensis, Timor Python, **47**
Pythoninae, 8
Pythonodipsas carinata, Keeled Snake, **370**
quadrangularis, Gyalopion, Desert Hooknose Snake, **338**
Quarantine, 650
Queen Snake, *Regina septemvittata,* **276**
Queretaro Ground Snake, *Conopsis lineatus,* **75**
Racer, *Coluber* sp., **599**
raddei, Vipera, Radde's Viper, **509**

729

Radde's Viper, *Vipera raddei*, **509**
Radiant Ground Snake, *Sonora aemula*, **374**
radiata, Elaphe, Stripe-tailed Rat Snake, **122**
radix, Thamnophis, Plains Garter Snake, **288, 293, 614**
Rainbow Boa, *Epicrates cenchria*, **55**
Rainbow Snake, *Farancia erytrogramma erytrogramma*, **239, 240, 241**
Ramphotyphlops nigrescens, Australian Blind Worm Snake, **33**
ramsayi, Aspidites, Ramsay's Python, **39**
Ramsay's Python, *Aspidites ramsayi*, **39**
ravergieri, Coluber, Mountain Racer, **98, 100, 101**
Rayed Rat Snake, *Elaphe rufodorsata*, **125**
Rearing, of young, **684**
Records, keeping, **686**
Red Coachwhip, *Masticophis flagellum piceus*, **201, 203**
Red Diamond Rattlesnake, *Crotalus ruber*, **554–555**
Red Milk Snake, *Lampropeltis triangulum syspila*, **160, 187**
Red-banded Snake, *Dinodon rufozonatum*, **237**
Red-bellied Black Snake, *Pseudechis porphyriacus*, **444–445**
Red-headed Krait, *Bungarus flaviceps*, **407**
Red-headed Rat Snake, *Elaphe moellendorffi*, **14, 127, 128**
Red-naped Diadem Snake, *Furina diadema*, **414**
Red-sided Garter Snake, *Thamnophis sirtalis parietalis*, **285**
Red-spotted Diadem Snake, *Spalerosophis arenarius*, **230**
Red-spotted Garter Snake, *Thamnophis sirtalis concinnus*, **289**
Red-tailed Rat Snake, *Gonyosoma oxycephalum*, **260**
Redbelly Snake, *Storeria occipitomaculata*, **280, 281**
Redbelly Water Snake, *Nerodia erythrogaster erythrogaster*, **273**
Redstripe Ribbon Snake, *Thamnophis proximus rubrilineatus*, **292**
Regal Ringneck Snake, *Diadophis punctatus regalis*, **256**
Regina grahami, Graham's Crayfish Snake, **277**
Regina rigida rigida, Glossy Crayfish Snake, **276**
Regina septemvittata, Queen Snake, **276**
reginae, Liophis, Royal Ground Snake, **81**
regius, Python, Ball Python, **48, 612, 661, 669**
reinhardtii, Calabaria, Calabar Burrowing Python, **40, 638**
Reproductive system, **674**
Reptiles, as a food item, **639**
Respiration, **610**
Reticulated Python, *Python reticulatus*, **46, 662**
reticulatus, Python, Reticulated Python, **46,**

662
rhabdocephalus, Xenodon, False Fer-de-lance, **88**
Rhabdophis subminiatus, Malayan Bush Snake, **278**
Rhabdophis tigrina, Chinese Tiger Snake, **278**
Rhadinaea brevirostris, Amazonian Short-nosed Snake, **86**
Rhadinaea flavilata, Pine Woods Snake, **86**
Rhamphiophis oxyrhynchus, Hook-nosed Beaked Snake, **370, 371, 372**
Rhinobothryum bovalli, Costa Rican Tree Snake, **373**
Rhinoceros Viper, *Bitis gabonica rhinoceros*, **476**
Rhinoceros Viper, *Bitis nasicornis*, **476, 478**
Rhinocheilus lecontei antonii, Mexican Longnose Snake, **225**
Rhinocheilus lecontei tessellatus, Texas Longnose Snake, **224, 225**
Rhodesian Sand Snake, *Psammophis jallae*, **362**
rhodostoma, Agkistrodon, Malaysian Moccasin, **529**
rhombeatus, Causus, Rhombic Night Adder, **510, 512**
rhombeatus, Psammophylax, Gray-bellied Grass Snake, **367, 368**
Rhombic Night Adder, *Causus rhombeatus*, **510, 512**
rhombifer rhombifer, Nerodia, Diamondback Water Snake, **274**
rhombifer, Oxyrhopus, Argentinian Bush Racer, **357, 360**
Rhynchocalamus melanocephalus, Asian Black-headed Snake, **252**
rhynchops, Cerberus, Dog-faced Water Snake, **309**
ridgewayi, Lytorhynchus, Afghan Awl-headed Snake, **197**
Ridley's Stripe-tailed Rat Snake, *Elaphe taeniura ridleyi*, **121**
rigida rigida, Regina, Glossy Crayfish Snake, **276**
Ring-necked Coffee Snake, *Ninia sebae*, **249**
Ringed Cat-eyed Snake, *Leptodeira annulata*, **346**
Ringed Python, *Liasis boa*, **44**
Ringed Water Cobra, *Boulengerina annulata*, **403**
Ringneck Snake, *Diadophis punctatus*, **255, 256**
rosaliae, Elaphe [Bogertophis], Baja California Rat Snake, **125**
rothi, Eirenis, Roth's Dwarf Snake, **115**
Roth's Dwarf Snake, *Eirenis rothi*, **115**
Rough Green Snake, *Opheodrys aestivus*, **206**
Rough-scaled Bush Viper, *Atheris squamiger*, **24, 463, 464, 465, 466, 467, 470, 611**
Rough-skinned Ground Snake, *Stenorrhina freminvillei*, **382**
rowleyi, Bothrops, Rowley's Pit Viper, **542**
Rowley's Pit Viper, *Bothrops rowleyi*, **542**

Royal Black-striped Snake, *Coniophanes imperialis*, 324–325
Royal Cat-eyed Snake, *Dipsadoboa aulica*, 327
Royal Ground Snake, *Liophis reginae*, 81
Rubber Boa, *Charina bottae*, 68–69, 70
ruber, Crotalus, Red Diamond Rattlesnake, 554–555
rubra cucullata, Tantilla, Blackhood Snake, 383
rufodorsata, Elaphe, Rayed Rat Snake, 125
rufozonatum, Dinodon, Red-banded Snake, 237
russelli, Vipera [Daboia], Russell's Viper, 491, 500–501, 508, 642
Russell's Cantil, *Agkistrodon bilineatus russeolus,* 525
Russell's Viper, *Vipera [Daboia] russelli,* 491, 500–501, 508, 642
rustica, Clelia, Smooth Brown Mussurana, 19, 322
ruthveni, Lampropeltis, Ruthven's Kingsnake, 177
ruthveni, Pituophis, Louisiana Pine Snake, 213
Ruthven's Kingsnake, *Lampropeltis ruthveni,* 177
Saddled Leafnose Snake, *Phyllorhynchus browni,* 209
Salt Marsh Snake, *Nerodia clarkii,* 270, 273
Salvadora deserticola, Big Bend Patchnose Snake, 227, 228–229
Salvadora grahamiae grahamiae, Mountain Patchnose Snake, 226, 227
Salvadora mexicanum, Mexican Patchnose Snake, 226
San Bernardino Mountain Kingsnake, *Lampropeltis zonata parvirubra,* 190
San Diego Mountain Kingsnake, *Lampropeltis zonata pulchra,* 191
San Francisco Garter Snake, *Thamnophis sirtalis tetrataenia,* 285, 299
San Luis Potosi Kingsnake, *Lampropeltis mexicana,* 169
San Pedro Mountain Kingsnake, *Lampropeltis zonata agalmae,* 188
Sand Snake, *Psammophis* sp., 363
Sanzania madagascariensis, Madagascar Tree Boa, 63
Sapera, *Xenodon merremi,* 88
sauritus nitae, Thamnophis, Bluestripe Ribbon Snake, 284, 301
sauritus sackeni, Thamnophis, Peninsula Ribbon Snake, 296–297
sauteri, Calliophis, Sauter's Oriental Coral Snake, 409
Sauter's Oriental Coral Snake, *Calliophis sauteri,* 409
Saw-scaled Viper, *Echis carinatus,* 487, 488, 489
sayi, Pituophis, Bullsnake, 215, 218, 680–681
scabra, Dasypeltis, Common Egg-eating Snake, 235, 641

scalaris, Elaphe, Ladder Snake, 124
Scarlet Kingsnake, *Lampropeltis triangulum elapsoides,* 157, 168, 669, 691
Scarlet Snake, *Cemophora coccinea,* 13, 91, 92
schistosa, Enhydrina, Beaked Sea Snake, 458
schlegeli, Bothrops, Eyelash Viper, 532, 533, 536, 537, 541, 543, 544, 545, 547, 645, 656
schlegeli, Typhlops, East African Blind Worm Snake, 33
schneideri, Bitis, Schneider's Puff Adder, 484
Schneider's Puff Adder, *Bitis schneideri,* 484
Schott's Whipsnake, *Masticophis taeniatus schotti,* 202
schrenckii, Elaphe, Manchurian Black Rat Snake, 124
Science, snakes in, 706
scutatus, Aspidelaps, Shield-nosed Cobra, 400, 401
scutatus, Notechis, Tiger Snake, 435, 436
scutellatus, Oxyuranus, Taipan, 440, 441
scytale, Anilius, False Coral Snake, 36–37
Sea Krait, *Laticauda* sp., 451, 452
sebae, Ninia, Ring-necked Coffee Snake, 249
semiannulata blanchardi, Sonora, Blanchard's Ground Snake, 378
semiannulata, Sonora, Great Plains Ground Snake, 375, 376–377, 379
semiannulata, Sonora, Ground Snake, 378, 379
semiannulatus, Telescopus, Half-ringed Cat Snake, 384, 385
semicinctus, Lystrophis, Half-Banded Hognose Snake, 83, 84
semifasciata, Laticauda, Half-banded Sea Krait, 452
Seminatrix pygaea cyclas, South Florida Swamp Snake, 279
Seminatrix pygaea, Black Swamp Snake, 279
semivariegatus, Philothamnus, Spotted Bush Snake, 208
semperi, Hydrophis, Lake Taal Banded Sea Snake, 459
Senses, of snakes, 598
septemvittata, Regina, Queen Snake, 276
septentrionalis septentrionalis, Leptodeira, Northern Cat-eyed Snake, 344–345, 348
septentrionalis, Leptodeira, Cat-eyed Snake, 347
Sex determination, 670
Sexing probe, 675
Sexing probe, ball-tip of, 675
sexlineatus, Liopholidophis, Six-lined Snake, 244
Sharp-tailed Snake, *Contia tenuis,* 105
Shaw's Two-colored Wolf Snake, *Lycodon striatus bicolor,* 245
Shedding, 592
Shedding, problems with, 665
Shield-nosed Cobra, *Aspidelaps scutatus,* 400, 401

731

Short-nosed Vine Snake, *Oxybelis brevirostris*, 353
Short-tailed Snake, *Stilosoma extenuatum*, 232–233
Short-tailed Snake, *Unechis brevicaudus*, 447
Short-tailed Viper, *Agkistrodon blomhoffi brevicaudis*, 526, 527
Shorthead Garter Snake, *Thamnophis brachystoma*, 287
Showmen, 704
Sibon nebulata, Central American Snail-eater, 393
Sibynophinae, 11
Sierra Mountain Kingsnake, *Lampropeltis zonata multicincta*, 189
signata, Hemiaspis, Black-bellied Swamp Snake, 416
Silver Vine Snake, *Oxybelis argenteus*, 354–355
simus, Heterodon, Southern Hognose Snake, 77
Sinaloan Milk Snake, *Lampropeltis triangulum sinaloae*, 158, 159, 185, 615
sipedon, Nerodia, Northern Water Snake, 628, 637
sirtalis concinnus, Thamnophis, Red-spotted Garter Snake, 289
sirtalis parietalis, Thamnophis, Red-sided Garter Snake, 285
sirtalis similis, Thamnophis, Bluestripe Garter Snake, 291, 294
sirtalis sirtalis, Thamnophis, Eastern Garter Snake, 288, 290, 298
sirtalis tetrataenia, Thamnophis, San Francisco Garter Snake, 285, 299
sirtalis, Thamnophis, Common Garter Snake, 286, 293, 299
Sistrurus catenatus catenatus, Eastern Massasauga, 562
Sistrurus catenatus, Massasauga, 561
Sistrurus miliarius miliarius, Carolina Pygmy Rattlesnake, 562
Sistrurus miliarius, Pygmy Rattlesnake, 563, 600
Six-lined Snake, *Liopholidophis sexlineatus*, 244
Skeleton, of cobra, 578
Skeleton, of pitviper, 588
Slender False Coral Snake, *Erythrolamprus bizonus*, 332–333
Smell, sense of, 601
smithi, Grayia, Smith's Water Snake, 260
Smith's Milk Snake, *Lampropeltis triangulum smithi*, 186
Smith's Water Snake, *Grayia smithi*, 260
Smooth Brown Mussurana, *Clelia rustica*, 19, 322
Smooth Earth Snake, *Virginia valeriae*, 307
Smooth Green Snake, *Opheodrys vernalis*, 205, 207
Smooth Ground Snake, *Liophis epinephalus*, 82
Smooth Snake, *Coronella austriaca*, 105, 106
Snail-sucker, *Dipsas* sp., 392
Snake charmer, 703
Snake charming, 702
Snake or lizard, distinction, 589
Societies, 705
Soft-banded Sea Snake, *Hydrophis inornatus*, 460
Solenoglyphic, 605
Solomon Islands Ground Boa, *Candoia carinata paulsoni*, 66
Sonora aemula, Radiant Ground Snake, 374
Sonora michoacanensis, Michoacan Ground Snake, 380–381
Sonora semiannulata blanchardi, Blanchard's Ground Snake, 378
Sonora semiannulata, Great Plains Ground Snake, 375, 376–377, 379
Sonora semiannulata, Ground Snake, 378, 379
Sonoran Gopher Snake, *Pituophis catenifer affinis*, 215, 631
Sonoran Lyre Snake, *Trimorphodon biscutatus lambda*, 390
Sonoran Shovelnose Snake, *Chionactis palarostris*, 96, 97
Sonoran Whipsnake, *Masticophis bilineatus*, 198, 204
South African Egg-eating Snake, *Dasypeltis inornata*, 236, 603
South American Ground Snake, *Liophis* sp., 81
South Florida Mole Kingsnake, *Lampropeltis calligaster occipitolineata*, 167
South Florida Swamp Snake, *Seminatrix pygaea cyclas*, 279
Southern Black Racer, *Coluber constrictor priapus*, 104
Southern Dwarf Garter Snake, *Homoroselaps lacteus*, 397
Southern Hognose Snake, *Heterodon simus*, 77
Southern Ringneck Snake, *Diadophis punctatus punctatus*, 257
Southern Smooth Snake, *Coronella girondica*, 106
Southwestern Speckled Rattlesnake, *Crotalus mitchelli pyrrhus*, 550
sp., *Aipysurus*, Olive Sea Snake, 455
sp., *Amphiesma*, Keeled Water Snake, 253
sp., *Coluber*, racer, 599
sp., *Dipsas*, Snail-sucker, 392
sp., *Helicops*, Keelback Snake, 262–263
sp., *Hydrophis*, banded sea snake, 597
sp., *Imantodes*, Tree Snake, 343
sp., *Laticauda*, Sea Krait, 451, 452
sp., *Liophis*, South American Ground Snake, 81
sp., *Pliocercus*, Big-scaled False Coral Snake, 275
sp., *Psammophis*, Sand Snake, 363
Spalerosophis arenarius, Red-spotted Diadem

732

Snake, 230
Spalerosophis diadema, Diadem Rat Snake, 230, 677
Speckled Kingsnake, *Lampropeltis getula holbrooki,* 173
Speckled Rattlesnake, *Crotalus mitchelli,* 673
Speed, of snakes, 597
Spilotes pullatus, Black and Yellow Rat Snake, 231
spilotes variegata, Morelia, Carpet Python, 45
Spitting Cobra, *Hemachatus haemachatus,* 415, 416
Spitting Cobra, *Naja mossambica,* 433
Spotted Adder, *Causus maculatus,* 511
Spotted Black Snake, *Pseudechis guttatus,* 441
Spotted Bush Snake, *Philothamnus semivariegatus,* 208
Spotted House Snake, *Lamprophis guttatus,* 243
Spotted Leafnose Snake, *Phyllorhynchus decurtatus,* 210
squamiger, Atheris, Rough-scaled Bush Viper, 24, 463, 464, 465, 466, 467, 470, 611
squamulosus, Cacophis, Golden-crowned Snake, 408
Sri Lankan Sand Boa, *Eryx colubrinus loveridgei,* 72
St. Helena Mountain Kingsnake, *Lampropeltis zonata zonata,* 191
stejnegeri, Trimeresurus, Stejneger's Pit Viper, 564
Stejneger's Pit Viper, *Trimeresurus stejnegeri,* 564
Stenorrhina freminvillei, Rough-skinned Ground Snake, 382
Steppe Ribbon Snake, *Psammophis lineolatus,* 365
Stilosoma extenuatum, Short-tailed Snake, 232–233
Stinking Goddess, *Elaphe carinata,* 134, 136–137
Stoke's Sea Snake, *Disteira [Astrotia] stokesi,* 455, 456
stokesi, Disteira [Astrotia], Stoke's Sea Snake, 455, 456
stolata, Amphiesma, New Guinean Keeled Water Snake, 253, 586
Storeria dekayi victa, Florida Brown Snake, 282–283
Storeria dekayi, Brown Snake, 280, 281
Storeria occipitomaculata, Redbelly Snake, 280, 281
Strategies, of feeding, 643
Streaked Kukri Snake, *Oligodon taeniolatus,* 250–251
streckeri, Ficimia, Mexican Hooknose Snake, 337
Stress, 663
striatus bicolor, Lycodon, Shaw's Two-colored Wolf Snake, 245
strigatus, Thamnodynastes, Amazonian Large-eyed Snake, 387
Stripe-tailed Rat Snake, *Elaphe radiata,* 122
Stuart's Milk Snake, *Lampropeltis triangulum stuarti,* 186, 577
stumpffi, Liopholidophis, Stumpff's Grass Snake, 244
Stumpff's Grass Snake, *Liopholidophis stumpffi,* 244
subcinctus, Lycodon, Banded Wolf Snake, 246
subflavus, Epicrates, Jamaican Boa, 55
subminiatus, Rhabdophis, Malayan Bush Snake, 278
subocularis, Elaphe [Bogertophis], Trans-Pecos Rat Snake, 116, 122, 685
Substrates, 630
sulphureus, Pseustes, Brazilian Liar, 221
Sumatran Cobra, *Naja naja sumatrana,* 431
Sunbeam Snake, *Xenopeltis unicolor,* 73
sundevalli boulengeri, Elapsoidea, Sundevall's Garter Snake, 414
Sundevall's Garter Snake, *Elapsoidea sundevalli boulengeri,* 414
Sunlight, 625
Sunrise Parrot Snake, *Leptophis diplotrophis,* 194
superbus, Austrelaps, Australian Copperhead, 402
superciliaris, Atheris, Lowland Viper, 470
Suta suta, Curl Snake, 446
suta, Suta, Curl Snake, 446
Systems, internal, 602
taeniatus schotti, Masticophis, Schott's Whipsnake, 202
taeniatus, Oligodon, Variegated Kukri Snake, 249
taeniolatus, Oligodon, Streaked Kukri Snake, 250–251
taeniura ridleyi, Elaphe, Ridley's Stripe-tailed Rat Snake, 121
taeniura yunnanensis, Elaphe, Yunnan Green Rat Snake, 135
Taipan, *Oxyuranus scutellatus,* 440, 441
Tan Tree Snake, *Boiga fusca,* 318
Tanganyikan Two-headed Snake, *Chilorhinophis gerardi,* 397
Tank, setup, 615, 617, 631
Tantilla nigriceps, Plains Blackheaded Snake, 382, 383
Tantilla rubra cucullata, Blackhood Snake, 383
tataricus, Eryx, Giant Sand Boa, 71
Taylor's Cantil, *Agkistrodon bilineatus taylori,* 513, 521
Teeth, 604
Telescopus beetzii, Namib Tiger Snake, 386
Telescopus dhara, Large-eyed Snake, 384
Telescopus fallax, False Cat Snake, 386
Telescopus semiannulatus, Half-ringed Cat Snake, 384, 385
Tentacled Snake, *Erpeton tentaculatus,* 311
tentaculatus, Erpeton, Tentacled Snake, 311
tenuis, Contia, Sharp-tailed Snake, 105
Terraria, built-in, 620

733

Terraria, fiberglass, 618
Terraria, glass, 616
Terraria, timber, 618
Tesselated Water Snake, *Natrix tessellata*, 264, 268
tessellata, Natrix, Tesselated Water Snake, 264, 268
Texas Blind Snake, *Leptotyphlops dulcis,* 7
Texas Coral Snake, *Micrurus fulvius tenere,* 418, 424
Texas Indigo Snake, *Drymarchon corais erebennus,* 109
Texas Lined Snake, *Tropiclonion lineatum texanum,* 302–303
Texas Longnose Snake, *Rhinocheilus lecontei tessellatus,* 224, 225
Texas Lyre Snake, *Trimorphodon biscutatus vilkinsoni,* 389
Texas Rat Snake, *Elaphe obsoleta lindheimerii,* 118, 119, 127
Texas Rat/Yellow Rat crossbreed, *Elaphe obsoleta lindheimerii x quadrivittata,* 118
textilis, Pseudonaja, Eastern Brown Snake, 446
Thamnodynastes strigatus, Amazonian Large-eyed Snake, 387
Thamnophis brachystoma, Shorthead Garter Snake, 287
Thamnophis butleri, Butler's Garter Snake, 284
Thamnophis cyrtopsis, Blackneck Garter Snake, 301
Thamnophis eques megalops, Mexican Garter Snake, 300
Thamnophis hammondii, Two-striped Garter Snake, 294
Thamnophis marcianus, Checkered Garter Snake, 289, 295
Thamnophis ordinoides, Northwestern Garter Snake, 300
Thamnophis proximus proximus, Western Ribbon Snake, 292
Thamnophis proximus rubrilineatus, Redstripe Ribbon Snake, 292
Thamnophis radix, Plains Garter Snake, 288, 293, 614
Thamnophis sauritus nitae, Bluestripe Ribbon Snake, 284, 301
Thamnophis sauritus sackeni, Peninsula Ribbon Snake, 296–297
Thamnophis sirtalis concinnus, Red-spotted Garter Snake, 289
Thamnophis sirtalis parietalis, Red-sided Garter Snake, 285
Thamnophis sirtalis similis, Bluestripe Garter Snake, 291, 294
Thamnophis sirtalis sirtalis, Eastern Garter Snake, 288, 290, 298
Thamnophis sirtalis tetrataenia, San Francisco Garter Snake, 285, 299
Thamnophis sirtalis, Common Garter Snake, 286, 293, 299
Thelotornis kirtlandi capensis, Cape Twig Snake, 387
Tick, removal of, 661, 667
Ticks, 666
Tiger Snake, *Notechis scutatus,* 435, 436
tigrina, Rhabdophis, Chinese Tiger Snake, 278
Timber Rattlesnake, *Crotalus horridus horridus,* 5, 559, 699
Timor Python, *Python timorensis,* 47
timorensis, Python, Timor Python, 47
Todos Santos Island Kingsnake, *Lampropeltis zonata herrerae,* 189
Tomodon dorsatus, Brown-spotted Night Viper, 388
Tomodon ocellatus, Ocellated Night Viper, 388
torquata, Hypsiglena, Night Snake, 341
Touch, sense of, 601
Trans-Pecos Rat Snake, *Elaphe [Bogertophis] subocularis,* 116, 122, 685
Tree Snake, *Imantodes* sp., 343
triangulum abnorma, Lampropeltis, Guatemalan Milk Snake, 177
triangulum amaura, Lampropeltis, Louisiana Milk Snake, 178
triangulum andesiana, Lampropeltis, Andean Milk Snake, 172
triangulum annulata, Lampropeltis, Mexican Milk Snake, 153, 154–155, 178
triangulum arcifera, Lampropeltis, Jalisco Milk Snake, 163, 179
triangulum blanchardi, Lampropeltis, Blanchard's Milk Snake, 179
triangulum campbelli, Lampropeltis, Pueblan Milk Snake, 153, 161, 180
triangulum celaenops, Lampropeltis, New Mexico Milk Snake, 180
triangulum conanti, Lampropeltis, Conant's Milk Snake, 161, 181
triangulum dixoni, Lampropeltis, Dixon's Milk Snake, 181
triangulum elapsoides, Lampropeltis, Scarlet Kingsnake, 157, 168, 669, 691
triangulum gaigeae, Lampropeltis, Black Milk Snake, 182
triangulum gentilis, Lampropeltis, Central Plains Milk Snake, 182
triangulum hondurensis, Lampropeltis, Honduran Milk Snake, 162, 183
triangulum micropholis, Lampropeltis, Ecuadorian Milk Snake, 183
triangulum multistrata, Lampropeltis, Pale Milk Snake, 184
triangulum nelsoni, Lampropeltis, Nelson's Milk Snake, 173
triangulum oligozona, Lampropeltis, Pacific Central American Milk Snake, 184
triangulum polyzona, Lampropeltis, Atlantic Central American Milk Snake, 185
triangulum sinaloae, Lampropeltis, Sinaloan Milk Snake, 158, 159, 185, 615
triangulum smithi, Lampropeltis, Smith's Milk Snake, 186
triangulum stuarti, Lampropeltis, Stuart's Milk

Snake, **186, 577**
triangulum syspila, Lampropeltis, Red Milk Snake, **160, 187**
triangulum taylori, Lampropeltis, Utah Milk Snake, **187**
triangulum "temporalis", Lampropeltis, Coastal Plains Milk Snake, **156**
triangulum triangulum, Lampropeltis, Eastern Milk Snake, **152, 188, 658**
triaspis mutabilis, Elaphe [Senticolis], Mexican Green Snake, **121**
triaspis, Elaphe [Senticolis], Green Rat Snake, **116, 135**
trigeminus, Oxyrhopus, Brazilian Bush Racer, **360**
Trimeresurus albolabris, White-lipped Tree Viper, **569**
Trimeresurus gramineus, Bamboo Pit Viper, **566**
Trimeresurus macgregori, MacGregor's Pit Viper, **566**
Trimeresurus popeorum, Pope's Pit Viper, **567, 568**
Trimeresurus puniceus, Flat-nosed Pit Viper, **570**
Trimeresurus purpureomaculatus, Mangrove Pit Viper, **564, 565, 568**
Trimeresurus stejnegeri, Stejneger's Pit Viper, **564**
Trimeresurus wagleri, Wagler's Pit Viper, **567, 569, 570, 571**
Trimorphodon biscutatus biscutatus, Common Lyre Snake, **390**
Trimorphodon biscutatus lambda, Sonoran Lyre Snake, **390**
Trimorphodon biscutatus vandenburghi, California Lyre Snake, **389**
Trimorphodon biscutatus vilkinsoni, Texas Lyre Snake, **389**
Trimorphodon biscutatus, Lyre Snake, **627**
Trinket Snake, *Elaphe helenae,* **134**
tritaeniatus, Psammophylax, White-bellied Grass Snake, **17, 368, 369**
trivirgata gracia, Lichanura, Desert Rosy Boa, **61**
trivirgata "myriolepis", Lichanura, Central Baja Rosy Boa, **62**
trivirgata trivirgata, Lichanura, Mexican Rosy Boa, **61, 62**
Tropical Cantil, *Agkistrodon bilineatus,* **513, 514, 518, 519, 524, 526**
Tropiclonion lineatum lineatum, Common Lined Snake, **304–305**
Tropiclonion lineatum texanum, Texas Lined Snake, **302–303**
Tropidophis canus curtus, Bimini Island Dwarf Boa, **64**
Tropidophis melanurus melanurus, Cuban Black-tailed Dwarf Boa, **64**
Tucson Shovelnose Snake, *Chionactis occipitalis klauberi,* **97**
Turkish Sand Boa, *Eryx jaculus turcicus,* **72**

Two-lined Burrowing Snake, *Elapomorphus bilineatus,* **330**
Two-striped Garter Snake, *Thamnophis hammondii,* **294**
Two-striped Shovel-nosed Snake, *Prosymna bivittata,* **219**
Typhlopidae, **6**
Typhlops schlegeli, East African Blind Worm Snake, **33**
typus, Dispholidus, Boomslang, **328, 329, 330**
Unechis brevicaudus, Short-tailed Snake, **447**
Unechis gouldi, Black-headed Snake, **448**
Unechis nigrostriatus, Black-striped Snake, **447**
Ungaliophis continentalis, Chiapan Boa, **65**
unicolor, Xenopeltis, Sunbeam Snake, **73**
Urodeum, **609**
Uromacer dorsalis, Haitian Longtail Snake, **87**
Uropeltidae, **10**
Urutu, *Bothrops alternatus,* **534**
Utah Milk Snake, *Lampropeltis triangulum taylori,* **187**
Utah Mountain Kingsnake, *Lampropeltis pyromelana infralabialis,* **175**
valeriae elegans, Virginia, Western Earth Snake, **306**
valeriae, Virginia, Smooth Earth Snake, **307**
Variegated Kukri Snake, *Oligodon taeniatus,* **249**
Venezuelan Tropical Racer, *Mastigodryas bifossatus,* **204**
Ventiliation, **629**
Vermicella annulata, Bandy-bandy, **448, 449**
vernalis, Opheodrys, Smooth Green Snake, **205, 207**
Vipera ammodytes transcaucasiana, Caucasus Sand Viper, **509**
Vipera aspis, European Asp, **492–493**
Vipera berus, Northern Viper, **23, 498–499, 504, 505, 508**
Vipera bornmulleri, Bornmuller's Viper, **506**
Vipera [Daboia] russelli, Russell's Viper, **491, 500–501, 508, 642**
Vipera kaznakovi, Caucasus Viper, **505, 506**
Vipera latifi, Latefi's Viper, **507**
Vipera lebetina, Levantine Viper, **495, 507**
Vipera palaestinae, Palestine Viper, **502–503**
Vipera [Pseudocerastes] fieldi, Israeli Horned Viper, **477**
Vipera raddei, Radde's Viper, **509**
Vipera xanthina, Ottoman Viper, **494, 496–497**
vipera, Cerastes, Common Sand Viper, **485**
Viperidae, **22**
Viperinae, **22**
Viperine Water Snake, *Natrix maura,* **265, 266–267**
Virgin Islands Tree Boa, *Epicrates monensis granti,* **56**
Virginia valeriae elegans, Western Earth Snake, **306**
Virginia valeriae, Smooth Earth Snake, **307**
viridis oreganus, Crotalus, Northern Pacific

Rattlesnake, 549, 557
viridis, Chondropython, Green Tree Python, 41, 42
vudi picticeps, Alsophis, Bimini Racer, 74
vulpina vulpina, Elaphe, Western Fox Snake, 120
wagleri, Trimeresurus, Wagler's Pit Viper, 567, 569, 570, 571
Wagler's Pit Viper, *Trimeresurus wagleri,* 567, 569, 570, 571
Walterinnesia aegyptia, Egyptian Desert Cobra, 449, 450
Western Blind Snake, *Leptotyphlops humilis,* 35
Western Coachwhip, *Masticophis flagellum testaceus,* 200, 202
Western Cottonmouth, *Agkistrodon piscivorus leucostoma,* 516–517
Western Diamondback Rattlesnake, *Crotalus atrox,* 551, 552–553, 556
Western Earth Snake, *Virginia valeriae elegans,* 306
Western Fox Snake, *Elaphe vulpina vulpina,* 120
Western Hognosed Snake, *Heterodon nasicus,* 76, 593, 607
Western Hooknose Snake, *Gyalopion canum,* 338
Western Leafnose Snake, *Phyllorhynchus decurtatus perkinsi,* 208
Western Ribbon Snake, *Thamnophis proximus proximus,* 292
Western Shovelnose Snake, *Chionactis occipitalis,* 96
Western Yellowbelly Racer, *Coluber constrictor mormon,* 103
White-bellied Grass Snake, *Psammophylax tritaeniatus,* 17, 368, 369
White-lipped Snake, *Drysdalia coronoides,* 413
White-lipped Tree Viper, *Trimeresurus albolabris,* 569
White-lipped Water Snake, *Lycodonomorphus whytii,* 247
White-spotted Slug Snake, *Pareas margaritophorus,* 394, 395
whytii, Lycodonomorphus, White-lipped Water Snake, 247
Wound, with sutures, 662
Wounds, 663
xanthina, Vipera, Ottoman Viper, 494, 496–497
Xenocalamus mechovi inornatus, Ornate Quill-nosed Snake, 391
Xenochrophis piscator, Checkered Keelback, 308
Xenoderminae, 11
Xenodon merremi, Sapera, 88
Xenodon rhabdocephalus, False Fer-de-lance, 88
Xenodontinae, 11
Xenopeltidae, 10
Xenopeltis unicolor, Sunbeam Snake, 73
xeropaga, Bitis, Desert Mountain Adder, 483
Yellow Anaconda, *Eunectes notaeus,* 60
Yellow Grass Snake, *Natrix natrix helvetica,* 268
Yellow Rat Snake, *Elaphe obsoleta quadrivittata,* 123, 126, 641
Yellow-bellied Sea Snake, *Pelamis platurus,* 460, 461
Yellow-bellied Water Snake, *Enhydris plumbea,* 310
Yellow-faced Whip Snake, *Demansia psammophis,* 411
Yellow-headed Mussurana, *Clelia occipitolutea,* 322
Yellow-jawed Lancehead, *Bothrops asper,* 535, 537
Yellow-lipped Palm Viper, *Bothrops laterlis,* 543, 545
Yellow-lipped Sea Krait, *Laticauda colubrina,* 451, 453
Yellow-tailed Indigo Snake, *Drymarchon corais corais,* 110
Yellowbelly Water Snake, *Nerodia erythrogaster flavigaster,* 14, 269, 272
Yunnan Green Rat Snake, *Elaphe taeniura yunnanensis,* 135
Zaocys dhumnades, Big-eye Snake, 234
zonata agalmae, Lampropeltis, San Pedro Mountain Kingsnake, 188
zonata herrerae, Lampropeltis, Todos Santos Island Kingsnake, 189
zonata multicincta, Lampropeltis, Sierra Mountain Kingsnake, 189
zonata multifasciata, Lampropeltis, Coastal Mountain Kingsnake, 190, 192
zonata parvirubra, Lampropeltis, San Bernardino Mountain Kingsnake, 190
zonata pulchra, Lampropeltis, San Diego Mountain Kingsnake, 191
zonata zonata, Lampropeltis, St. Helena Mountain Kingsnake, 191
zonata, Lampropeltis, Mountain Kingsnake, 164–165, 192
Zoos, 705